TRANSPORTATION INFRASTRUCTURE - ROADS, HIGHWAYS, BRIDGES, AIRPORTS AND MASS TRANSIT

TRAFFIC BOTTLENECKS

CHALLENGES AND REDUCTION STRATEGIES

Transportation Infrastructure - Roads, Highways, Bridges, Airports and Mass Transit

Additional books in this series can be found on Nova's website under the Series tab.

Additional E-books in this series can be found on Nova's website under the E-book tab.

Transportation Issues, Policies and R&D

Additional books in this series can be found on Nova's website under the Series tab.

Additional E-books in this series can be found on Nova's website under the E-book tab.

TRANSPORTATION INFRASTRUCTURE - ROADS, HIGHWAYS, BRIDGES, AIRPORTS AND MASS TRANSIT

TRAFFIC BOTTLENECKS

CHALLENGES AND REDUCTION STRATEGIES

ALEX L. SLOAN
AND
CONNIE F. DORSEY
EDITORS

Nova Science Publishers, Inc.
New York

For permission to use material from this book please contact us:
Telephone 631-231-7269; Fax 631-231-8175
Web Site: http://www.novapublishers.com

Additional color graphics may be available in the e-book version of this book.

Library of Congress Cataloging-in-Publication Data

ISBN: 978-1-62081-180-1

Published by Nova Science Publishers, Inc. † New York

CONTENTS

PREFACE

Much of recurring congestion is due to physical bottlenecks - potentially correctable points on the highway system where traffic flow is restricted. While many of the nation's bottlenecks can only be addressed through costly major construction projects or costly transportation alternative solutions (i.e., high occupancy vehicle or toll lanes, dynamic pricing, investments in transit alternatives, etc.), there is significant opportunity for the application of operational and low-cost "fixes" at spot-specific locations. This book provides guidance to state and local transportation personnel on how to overcome barriers and challenges with reference to the implementation of localized congestion relief projects.

Chapter 1 - The delays arising from traffic congestion are an unavoidably frustrating fact of life. Or are they – unavoidable, that is? Why must we accept to take thirty minutes to make what should be a fifteen minute drive? Speed is distance divided by time; unless of course one has to pointlessly add time sitting in bumper to bumper traffic! In that case, speed drifts into "delay," which in turn drifts into our lives. In bygone days one could expect a vehicle trip to be governed only by the "straight line" distance and the performance of car and driver (with a necessary nod to speed limits too).

In today's world, that same trip often factors "expectations" of congestion delay caused by others. Not us, mind you, but others who, if they would only get out of our way, would free that trip back to the bygone era.

Chapter 2 - The Federal Highway Administration's (FHWA) Localized Bottleneck Reduction Initiative (LBR) program focuses attention on mitigating the operational causes of recurring congestion "hot spots" (i.e., traffic bottlenecks) at ramps, merges, lane drops, intersections, weaves, etc. One of the efforts of the LBR program is to encourage agencies to adopt a defined, "named" annualized spot-congestion program in the same manner that they might have an annualized spot-safety program for high crash locations. In the course of conducting state visits to "spread the gospel" of the tremendous benefits of clearing up one or more congested bottlenecks, the LBR staff has occasionally heard "push back" from some agencies that cite institutional or other barriers to enacting either individual projects or agency-wide programs. Examples of some of these barriers would be "we can't enact (these types of solutions) because (we feel) they violate firm safety design tenets or regulations;" or, "we can't undertake a spot-solution on a freeway absent having a vetted, adopted, twenty-year plan (or similar) already in place;" or, "how would such projects affect (our) Metropolitan Planning Organization's (MPOs) air quality, nonattainment status?" Conversely, the LBR staff has conducted state visits wherein these questions never came up; either signifying no such concerns, or success in overcoming them.

Chapter 3 - FHWA's Localized Bottleneck Reduction Initiative (LBR) program has researched over the last five years the causes, impacts, and mitigations available to combat localized recurring congestion; that is, congestion that is primarily "point-specific" as to cause, location, duration, and repetitiveness. This guidance document is intended to aid agencies in establishing either ad hoc or annualized programs that address localized congestion, much in the same way that an annualized safety-spot program would address localized safety issues.

"Localized" recurring congestion differs from regional or corridor-sized congestion, in that the former is characterized by a relatively low-cost, usually correctable operational deficiency (e.g., lane drop, weave, merge, etc.), whereas the latter is often systemic, and may be sufficiently complex in cause and relief that funding and/or project duration are typically very high and very long, respectively.

An agency that does not have a stand-alone "localized" congestion program is missing an opportunity to address a specific subset of overall congestion.

In: Traffic Bottlenecks
Editors: A. L. Sloan and C. F. Dorsey
ISBN: 978-1-62081-180-1

Chapter 1

RECURRING TRAFFIC BOTTLENECKS: A PRIMER FOCUS ON LOW-COST OPERATIONAL IMPROVEMENTS*

United States Department of Transportation
Federal Highway Administration

INTRODUCTION

The delays arising from traffic congestion are an unavoidably frustrating fact of life. Or are they – unavoidable, that is? Why must we accept to take thirty minutes to make what should be a fifteen minute drive? Speed is distance divided by time; unless of course one has to pointlessly add time sitting in bumper to bumper traffic! In that case, speed drifts into "delay," which in turn drifts into our lives. In bygone days one could expect a vehicle trip to be governed only by the "straight line" distance and the performance of car and driver (with a necessary nod to speed limits too).

In today's world, that same trip often factors "expectations" of congestion delay caused by others. Not us, mind you, but others who, if they would only get out of our way, would free that trip back to the bygone era.

* This is an edited, reformatted and augmented version of United States Department of Transportation Federal Highway Administration, Publication, dated June 9.

According to a February 2007 Harris Poll, just over one-third (37 percent) of respondents cite traffic congestion as a serious problem in the community, while one-quarter say traffic congestion is a serious problem that is not being addressed. But please don't tell the thousands of practicing traffic engineers, planners, and road workers that congestion is not being addressed. It's just that much like weather forecasting, traffic management is a dynamic activity that makes it an ever evolving profession. And like weather forecasting, we are getting better and better at it, but remain at the whim of unrelenting and ever-evolving "fronts." But there are some conditions we can do something about given the right resolve.

> "Close to half of all congestion happens day after day at the same time and location."
>
> Source: http://www.fhwa.dot.gov/congestion/describing_problem.htm.

Much of recurring congestion is due to physical bottlenecks – potentially correctible points on the highway system where traffic flow is restricted. While many of the nation's bottlenecks can only be addressed through costly major construction projects, there is a significant opportunity for the application of operational and low-cost infrastructure solutions to bring about relief at these chokepoints.

This document, *Recurring Traffic Bottlenecks: A Primer – Focus on Low-Cost Operational Improvements*, describes such facility breakdowns and explores the opportunity for near-term operational and low-cost construction opportunities to correct them.

This Primer is intended to be a dynamic work-in-progress. As newer strategies and discourse become available they will be uploaded to the FHWA's "Bottleneck" web site http://www.ops.fhwa.dot.gov/bn/index.htm and the Primer will be periodically updated.

This Primer constitutes "Version 2." The Primer is a key resource for Federal Highway Administration's Localized Bottleneck Reduction (LBR) Program, providing a virtual forum for peer exchange between members of the transportation community interested in alleviating bottleneck congestion.

The LBR program, initiated in 2006, is designed to expand the portfolio of bottleneck reduction tools available to transportation agencies to encompass innovative, readily adopted strategies for reducing congestion at bottleneck locations.

UNDERSTANDING BOTTLENECKS

What is a "Traffic Bottleneck"?

Webster's dictionary defines a "bottleneck" as: i) a narrow or obstructed portion of a highway or pipeline, or ii) a hindrance to production or progress. Certainly the elemental characteristics of traffic bottlenecks exist in these descriptions. However, a road does not necessarily have to "narrow" for a bottleneck to exist (e.g., witness bottlenecks caused by a weave condition, sun glare, or a vertical climb).

Traffic bottlenecks (hereafter, bottlenecks) have a myriad of causes and durations. The most egregious ones tend to be freeway-to-freeway interchanges, but we all know that smaller, lesser chokepoints are frustrating too. Many of these choke-points are "operationally influenced bottlenecks," defined as localized sections of highway where traffic experiences reduced speeds and delays due to recurring operational conditions.

The fact that many recurring locations are "facility determinate" (i.e., the design condition contributes to the resulting backup) is both encouraging and discouraging. Facility design is a tangible feature that can always be improved; however the cost or the necessary right-of-way may be prohibitive.

Bottlenecks may be compared to constricted storm pipes that can carry only so much water – during floods the excess water just backs up, much the same as traffic does at bottleneck locations. Like the friction that results from turbulence in a pipe, once traffic flow breaks down to stop-and-go conditions, capacity is reduced – fewer cars can get through the bottleneck because of the extra turbulence. The options available to free up "the pipe" are to carry less water, increase the size of the pipe or system, or remove whatever constriction exists (i.e., remove the kink, reduce cavitations, or reduce head loss due to friction at junctures, et al.). Obviously the pipe comparison is analogous to options available to improve traffic flow; namely, to reduce demand (by effecting driver behavior), build new infrastructure, or improve that which already exists. Bottleneck mitigation is the third of these options.

What is "Congestion"?

FHWA's Traffic Congestion and Reliability Report defines congestion as *"an excess of vehicles on a roadway at a particular time resulting in speeds that are slower – sometimes much slower – than normal or free flow speeds; [Congestion is] stopand-go traffic."* The root causes of congestion have long been understood, and there is now broad consensus that congestion generally reflects a fundamental imbalance of supply and demand. That is, during hours of peak usage of the transportation facilities most desirable to motorists, the supply of roadway capacity is insufficient to meet the demand for those facilities.

Traffic congestion can be characterized as either "recurring" or "nonrecurring." This distinction is useful in helping the community of transportation professionals devise strategies that will either mitigate or reduce congestion.

Recurring congestion happens in roughly the same time and place on the same days of the week. It results when physical capacity is simply not adequate to accommodate demand during peak periods. Causes of recurring congestion include inadequate physical capacity and poor signal timing.

Nonrecurring congestion occurs randomly and is caused by events such as work zones, traffic incidents, and bad weather. Even if "planned" in some cases, these occurrences are irregular and are not predictably habitual or recurring in location. Obviously, when these nonrecurring events occur on an already congested facility, the impacts are magnified. Figure 1 presents a pie chart showing the factors that cause traffic congestion.

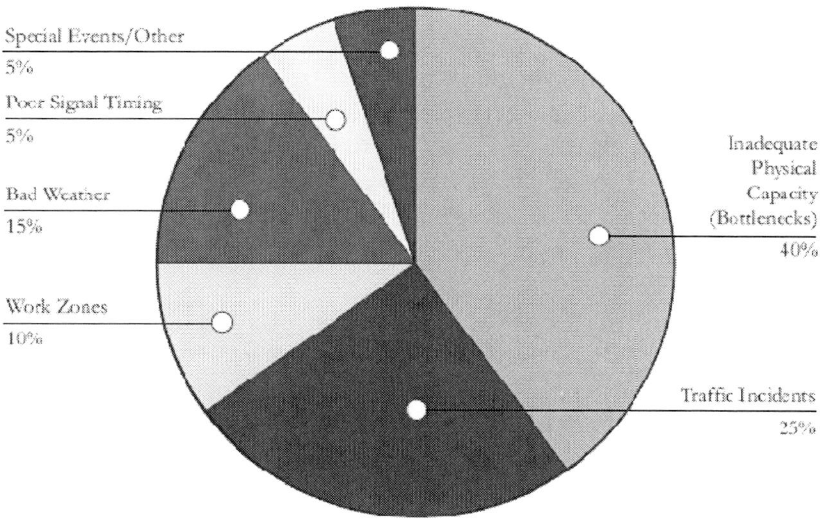

Source: http://www.fhwa.dot.gov/congestion/describing_problem.htm.

Figure 1. Sources of Traffic Congestion.

So Who's to Blame?

One side of a popular coin says, "There are not enough roads out there!" The other side says, "There are too many cars!" And the edge of the coin reads, "What's out there doesn't work well!" Boy, it's just like humans to blame everything else – except ourselves.

The title of Tom Vanderbilt's 2008 book "Traffic: Why We Drive the Way We Do (And What it Says About Us)" reflects the human traits – and not the physical network – that have conspired to result in the system that we have (and how we use it). Sure, there is the inevitable "state of the system" baseline that discusses the age of the network, some measures of effectiveness (or lack thereof) and a "where do we go from here" look forward. But an equal amount of pages are given to psychologists as to engineers in explaining why the need to overcome our human traits drives the incremental technological advancements; something akin to the old adage that building better and more roads begets buying more vehicles and making more trips. These human factors include driver attention, reaction, and tendencies, and explain that we are creatures of habit or need (witness our penchant to battle the peak hour despite our hatred for it), that we essentially drive "for our personal gain" instead of for the greater good (e.g., reference his comparisons of ants and other animal colonies that seem to work for a common goal), and that we suffer a myriad of transformations (e.g., road rage, selfishness, vehicle envy, distraction-bygadget, single-driver preference, et al.) that conspire to erode our driving skills and civility, the latter of which we are otherwise forced to address face-to-face when we are outside of our four-wheeled cocoons. Inside our cars we become emboldened; protected by 2,000 pounds of steel and anonymous behind our tinted windows.

> George Carlin famously observed that "everyone driving slower than you is an idiot and everyone faster is a maniac."

Any slight – in terms of delay, encroachment in our lane, or imposition on our perceived "right" to an unencumbered trip – is a personal affront. A driver passes us; we feel compelled to pass him. We switch to the "fast lane" and invariably so do others, such that we are now in the "slow lane." Vanderbilt cites research that demonstrates 1) how this creates "density friction" that degrades efficiency of traffic for everyone; 2) how you and a neighboring "spotter" car will progress roughly equal; and 3) the fact that all that jockeying gains nary a time advantage when weighed against the nominal case of just staying with one lane. But still we lane jump. Compare to the fact that if one flips a coin enough times, the odds of heads versus tails eventually narrows to 50 percent each. But we continue to switch lanes – or play the odds – thinking that we can beat them, because every once in a while, we do. Further evidence of our human consequence exists in the fact that most all agencies now use the term "crashes" and not "accidents" to record those incidents, to overcome the excuse that "accident" somehow removes responsibility from our conscience. It's not an "accident" (except perhaps to the wrong place/wrong time victim) if you had been paying due diligence, staying awake, not talking on your cell phone, or not driving while impaired.

Specific to the topic of congestion (and bottlenecks), Vanderbilt's citations confirm what we intrinsically know; that in survey after survey the majority of us consider ourselves superior drivers to practically anyone else despite that in other surveys we are very poor at estimating time spent in delay, distances traveled, and the cost of that delay. A quarter is worth twenty-five cents to everyone, but a minute is overestimated by the guy sitting in a queue, and underestimated by the guy racing to an appointment.

The aforementioned "superiority" manifests itself in our belief that somehow we are more worthy of being on the road and it is everyone else who is clogging it up. We are ridiculously inept at estimating our own speed ("I was going 30 MPH" when in fact he was clocked at 46); estimating time spent circling for a parking space ("I must have spent 5 minutes looking" when in fact she only spent 1 minute 45 seconds from street to space); and estimating others' speed ("everyone travels 50 down my street" when in fact, a radar study found an average of 34 MPH and only two cars doing above 46 MPH in a three-day study). Or maybe in the latter case, it was the same car observed twice; hard to tell, but a moot point in terms of accusing "everyone." We often travel at a speed that the road "allows" and not what the sign mandates. We're not intentionally flouting the law, but hey, we know better than the police and the engineers what it takes to get from here to there! After all, the engineers dutifully allow for a "safety factor" in their design; that's for me, right? Sometimes the prior examples of estimation can be written off as exaggeration but research consistently reproduces like results when people are observed first hand, or when focus groups are asked situational questions.

In the context of "a road never jumped up and bit me" it seems hardly right to "blame" the roads, absent poor maintenance or a malfunctioning signal. So we blame the planners, the engineers, the politicians, and the "idiots" on the road. Never mind that those same "idiots" are the neighbors, customers, and peers we somehow otherwise abide. Clearly our desire to drive more often, and further, and own more vehicles, has strained our major highway arteries. The effort becomes to reduce the *rate* of congestion, if not to eliminate it altogether in as many locations as funding and right-of-way can bear. All the while we will swim upstream against perpetual waves of drivers, mileage, and demand. So the battle against congestion wages on while our human instincts contribute to the malaise.

The Frustration of Congestion

It seems hardly necessary to state that we are frustrated by congestion. We are frustrated because we feel we can't control it. We can control most everything else in our lives. We control what, when, and how much TV we watch because we have the remote and the DVR! We control what, when, and how much we eat. To the best of our ability, we control what we do (our jobs) and where we do it; although we'd like a *little* more control on how much they pay us! We control our pastimes, our hobbies, and our vacations. We somewhat control our spouses and children; or at least we have influence upon them. But unlike the scene in *Bruce Almighty*, we can't part the "sea" of traffic in front of us like Jim Carrey did.

But do we have to *accept* congestion? Our grandparents – and theirs – didn't have to endure the levels of congestion that we see today. But they also didn't enjoy the extent of today's roadway network, the availability of gas stations and services, and the efficiencies and perks of modern cars. Time marches on, not back, so we are left to suffer the ravages of our success.

WHY CAN'T IT BE LIKE THIS ALL THE TIME? (UNCONGESTED, THAT IS!)

Shopping mall planners design parking lots for the twelfth highest demand day of the year so as not to "waste" valuable land that would otherwise be squandered 96 percent of the time. (Similar to insurance actuary tables, the "12th highest" axiom is an accord based on the intrinsic worth of land cost, parking demand, and of course, retail profit.) No doubt, during the Christmas shopping season, we all would love to get those "unnecessary" spaces back! Recurring congestion is sort of like that, except that highway builders don't intentionally under-build; it just seems that way. Highways are built to "design hour" and "design year" tenets, but population and traffic demand are ever growing – and rarely receding – over time. Outside of peak hours, the great majority of highway facilities are more than sufficient for the dynamic demand. Unfortunately, we tend to see highways at their busiest hours. But we occasionally experience the off-hour conditions too, i.e., when traffic flows freely. And like those too-few vacations when we are pampered and coddled, we ask "why can't it be like this all the time?" Recurring congestion is a function of physical constraints, as well as how *and when* drivers interact with a given facility and other vehicles on the facility.

A "Bottleneck" May Be "Congestion," but Congestion is Not Always "Just a Bottleneck"

The word "paper" can mean either a single sheet or a ream; likewise, "congestion" can be a single bottleneck – or much more. A bottleneck is distinguished from congestion in that it occurs at a specific location, and not pervasively along the entire corridor. Recognizing this difference is the first important distinction in developing strategies for one or the other. When too many vehicles compete along all segments of a facility, corridor- or systemwide "congestion" will inevitably result. It is overarching in nature. But when only subordinate segments of that facility are burdened, then operationally recurring bottlenecks are said to exist. In this context then, a bottleneck certainly constitutes "congestion," but congestion cannot be said to be universally analogous to a "bottleneck."

The FHWA estimates that 40 percent of all congestion nationwide can be attributed to recurring bottlenecks (i.e., inadequate physical capacity) and another 5 percent is attributable to inefficient traffic signalization. The good news is that all these things are potentially correctible by remediation. The bad news is that there are many, many candidate locations, and agencies are fiscally constrained on how much they can do. All things being equal, a recurring bottleneck will disappear once traffic demand has decreased to a point where the operational or geometric deficiency is no longer a factor, while a nonrecurring bottleneck will disappear only after the random event has been removed. A nonrecurring bottleneck may be further impacted by traffic volumes, but it is not caused by traffic demand. Conversely, only a physical improvement will relieve a recurring bottleneck.

What Elements Typically Exist to Define a "Bottleneck"?

The Localized Bottleneck Reduction Program focuses on operationally influenced locations; that is to say, those that have a fundamentally design-based cause, resulting in recurring delays of generally predictable times and durations. The root cause of traffic flow degradation at the subject point of a recurrent constriction is almost always a correctable problem. The following conditions either exist or help to identify a recurring bottleneck condition.

- **A traffic queue upstream of the bottleneck,** wherein speeds are below free-flow conditions elsewhere on the facility. (Note: if speeds at all or most-all of the facility are consistently and regularly lower than free-flow speeds, then overarching congestion exists. This is congestion beyond a mere point-specific bottleneck location.)
- **A beginning point for a queue.** There should be a definable point that separates upstream and downstream conditions. The geometry of that point is often coincidently the root cause of the operational deficiency.

- **Free flow traffic conditions downstream** of the bottleneck that have returned to nominal or design conditions.
- As it pertains to an operational deficiency, **a predictable recurring cause.** (Note: this implies that all things being equal, a solution exists that is nevertheless theoretically "correctable" by design, as opposed to, say, an amorphous, random event.)
- **Traffic volumes that exceed the capacity of the confluence to process traffic.** (Note: this applies to recurring events even more-so than nonrecurring.)

What Options Exist to Combat Congestion (and by Extension, Bottlenecks)?

Fixing operationally influenced deficiencies applies to the fourth of the following four strategies available to combat congestion.

- **Bring supply and demand in alignment through congestion pricing.** Congestion pricing or peak-period pricing entails fees or tolls for road use that varies by level of vehicle demand on the facility. As with market pricing in other sectors, road pricing helps allocate limited supply – in this case, that of available road space. With user charges assesses at the point of use, greater efficiency results through improved response to market forces. Charges are typically assessed electronically to eliminate delays associated with manual toll collection facilities. Road-use charges that vary with the level of vehicle demand provide incentives to shift some trips to off-peak times, less congested routes, or alternative modes; or to cause some lower-value trips to be combined with other trips or simply to be eliminated. Congestion pricing has several important objectives. First, it seeks to balance demand with available capacity, i.e., the supply of road space. Second, it seeks to fairly allocate the costs associated with operating, maintaining, and expanding the transportation system to meet growing demand. Third, it seeks to improve operation of the highway system. A fourth objective may include revenue generation.

- **Provide alternatives as to how, when, where, and if to travel.** The goal of this strategy is to reduce the number of vehicles on a given road. This may take the form of promoting alternative commute options such as employee telecommuting options or making transit easier and more attractive to use. Also of interest in managing demand are driver incentive programs that, for example, promote ridesharing and off-peak use.
- **Invest in new highway capacity.** Add new construction on new alignments to preserve or improve system performance.
- **Improve the management and operation of the system.** Improve the day-to-day operation of the system by retiming traffic signals, applying access management techniques, removing operational deficiencies, and improving response time and management of traffic disrupting events like work zones, accidents, and special events. Provide real time information about the system so that travelers can make immediate decisions about when, where, and how to travel, and transportation agencies can make real-time adjustments to improve system operations.

Why Do Recurring Bottlenecks Occur?

Every highway facility has decision points such as on and off ramps, merge areas, weave areas, lane drops, tollbooth areas, and traffic signals; or design constraints such as curves, climbs, underpasses, and narrow or nonexistent shoulders. In many thousands of cases, these operational junctions and characteristics operate sufficiently and anonymously; however, when the design itself becomes the constricting factor in processing traffic demand, then an operationally influenced bottleneck can result.

The degree of congestion at a bottleneck location is related to its physical design. Some operational junctions were constructed years ago using design standards now considered to be antiquated, while others were built to sufficiently high design standards but are simply overwhelmed by traffic demand. Whatever the root cause, operationally influenced bottlenecks can occur at:

- A **lane drop** particularly mid-segment where one or more traffic lanes are lost. These typically appear at bridge crossings and in work zones. The latter, however, is a nonrecurring event and is usually remedied when the work zone is removed. Ideally, lane drops should be located at exit ramps where there is a large volume of exiting traffic.
- A **weaving area**, where traffic must merge across one or more lanes to access entry or exit ramps. Bottleneck conditions are worsened when there are confusing or insufficient weaving lengths.
- **Freeway on-ramps,** which are merging areas where traffic from local streets can join a freeway. Bottleneck conditions are worsened on freeway on-ramps without auxiliary lanes, short acceleration ramps, or where there are multiple on-ramps in close proximity.
- **Freeway-to-freeway interchanges,** which are special cases of on-ramps where flow from one freeway is directed to another. These are typically the most severe form of physical bottlenecks because of the high traffic volumes involved.
- **Freeway exit ramps,** which are diverging areas where traffic leaves a freeway. Bottleneck conditions are worsened on freeway exit ramps that have a short ramp length, traffic signal deficiencies at the ramp terminal intersection, or other conditions that may cause ramp queues to back up onto freeway mainlanes.
- **Abrupt changes in highway alignment,** which occur at sharp curves and hills and cause drivers to slow down either because of safety concerns or because their vehicles cannot maintain speed on upgrades. Another example of this type of bottleneck is in work zones where lanes may be redirected or "shifted" during construction.
- **Low clearance structures** such as tunnels and underpasses. Drivers slow to use extra caution, or overload bypass routes. Even sufficiently tall clearances may impart problems. In one renowned case, the mere optical illusion of a bridge appearing lower than it really is causes drivers to slow down, resulting in recurring bottleneck conditions.

- **Lane narrowing,** caused by either narrow travel lanes or narrow or nonexistent shoulders.
- **Intended interruptions to traffic flow** are literally "traffic disruptions on purpose" that are sometimes necessary in order to manage system flow. Traffic signals, freeway ramp meters, and tollbooths can all contribute to capacity loss.

Figure 2 lists various types of freeway bottleneck causes.

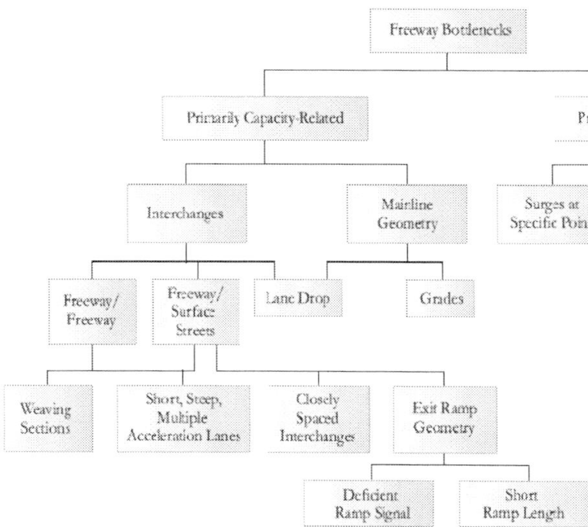

Figure 2. Types of Freeway Bottlenecks.

What is Stopping Us from Fixing Bottlenecks?

The knee jerk reaction might be "lack of money." But that's everyone's first complaint about, well, most every problem! In visiting with many states to ascertain if they have a bottleneck-specific program or similar that targets chokepoint congestion, we have found a sampling of reasons.

- **A predisposition to execute major projects.** Certainly, no one is faulting an agency which is attentive to executing major transportation initiatives. There is no shortage of "fronts" on which to fight the congestion battle. A short list would include HOVs, tolling and pricing, transit alternatives, ridesharing programs, and bridge rehabilitation. But the onerous processes involved in many of these initiatives can squeeze out smaller programs.

- **Lack of a champion.** Many successful state or metropolitan planning organization (MPO) programs are the result of one or more persons taking charge to either mandate or adopt a program.
  High-level administrators form the direction for their agencies. Mid-level managers' production reflects their priorities and skills in executing those initiatives.
- **Lack of a "named" program.** Unless there is an identity, bottleneck remediation is usually relegated to a few projects done "by rote," e.g., as part of an annualized safety program, or as a subordinate part of larger, other-purposed projects.
- **A culture of historical practices.** Example: an agency that dutifully executed an annualized "safety" program looked only at crash rates in determining their annual top ten list of projects.
  After instituting a congestion mapping process, they identified several significant stand-alone chokepoints that did not correlate with their high crash mapping. Thereafter, high *congestion* hot spots competed with high *accident* hot spots on their unified top ten list of projects.

Solving Recurring Bottlenecks is a Win-Win Situation for Nonrecurring Incidents Too!

Strategies to alleviate bottlenecks will lessen delay caused by nonrecurring events such as accidents, weather, work zones, etc. For example, consider an accident that blocks a single lane of traffic. If only two lanes existed prior to the accident occurring, the impact would be greater than if three lanes existed. The capacity loss resulting from the nonrecurring event will be lessened due to the improvements made to the system to benefit the recurring situation.

Common Myths about Bottlenecks

"Bottlenecks are caused only by not enough lanes on an extended highway section."

In the past, recurring congestion was thought to be an overarching or systemic problem (i.e., not enough lanes) that could only be resolved by widening an entire corridor. However, the fact that other subordinate sections *within the system* operate sufficiently demonstrates that the uniform highway segments may not necessarily be under-designed.

Traditional capital solutions often grew from the misconception that a multilane facility should be designed to alleviate recurring congestion during the peak hours each day.

The problem is that funding for such large scale projects is limited, and right-of-way is often restricted, such that these projects take many years to complete. As a result, recurring congestion historically goes untreated, or is forced to compete against other worthy projects, until funding becomes available to "catch up" to the problem that has grown from the day the facility opens.

With a shift in focus away from the perception that recurring congestion is systemic (and thus treatable with only large scale projects), it is possible to explore a wider range of improvement strategies that are possible in the short-term. While these will never replace the need for corridor-wide fixes – especially at "mega-bottlenecks" such as major freeway-to-freeway interchanges – low-cost, quick turnaround bottleneck fixes can provide congestion relief at the point of occurrence.

"Bottlenecks can't be fixed without massive reconstruction."

With the focus of traditional transportation planning and programming efforts on major capital projects, it has been assumed for many years that bottlenecks cannot be fixed without massive reconstruction of an interchange or corridor. However, there are numerous examples where agencies opted to make lower-cost improvements that resulted in significant improvements in traffic flow.

"Improving a bottleneck won't help traffic flow outside of peak periods."

Because traffic-influencing events like incidents, bad weather, work zones, and special events can happen at any time, congestion is not restricted to peak times of the day. Benefits realized from improvements made to address peak-period bottleneck problems will invariably carry over to the times outside of the peak when congestion occurs.

A Rogue's Gallery of Infamous Bottlenecks

On much of the urban highway system, there are specific points that are notorious for causing congestion on a daily basis. These locations – which can be a single interchange (usually freeway-to-freeway), a series of closely spaced interchanges, or lane drops – are focal points for congestion in corridors. Major bottlenecks tend to dominate the congestion problem in corridors where they exist, some even acquiring nicknames from local motorists such as the "Spaghetti Bowl," "Hillside Strangler," "Spaghetti Junction," "Malfunction Junction," and "Mixmaster".

Over the past several years, transportation professionals have come to realize that highway bottlenecks demand special attention. Several national studies have highlighted bottlenecks as a major congestion problem in urban areas. These studies have raised the level of awareness about bottlenecks as a problem, warranting that they be treated as a significant part of the congestion problem.

The American Highway Users Alliance (AHUA) conducted two studies of the nation's urban bottlenecks in 1999 and 2004. The studies produced rankings of the worst bottlenecks in terms of total delay to travelers and discussed what was being done to fix the problems at locations where specific improvements had been scheduled. The studies found that nearly all of the worst bottlenecks are major freeway-to-freeway interchanges in large urban areas. The 2004 study updated the rankings and discussed three bottleneck improvement success stories – bottlenecks identified in 1999 that were subsequently improved or well under construction.

FHWA's first effort related to bottlenecks was in the freight (trucking) arena. Using the AHUA studies as a starting point, the impact of bottlenecks on truck travel was assessed. Bottlenecks outside of urban areas were also considered (e.g., steep grades). A major finding of this study was that in terms of total delay, the urban bottlenecks – typically thought of as a commuter-related problem – are also major sources of truck delay.

States and regions are beginning to recognize the significance of bottlenecks as well. The Ohio Department of Transportation completed a study of freight (trucking) bottlenecks, and the Interstate-95 Corridor Coalition is undertaking a study of all potential bottlenecks in Coalition states. The Atlanta Regional Commission has defined bottlenecks as a specific portion of their Congestion Management Process and is identifying regional and local bottlenecks in their network.

Table 1. The Worst Physical Bottlenecks in the United States, 2008

2008 Rank	Area	Road/Direction	Segment/Interchange	State	Length	Hours Congested	Average Speed When Congested
1	New York	Cross Bronx Expressway/I-95 SB	Bronx River Parkway/Exit 48	NY	0.36	94	11.2
2	San Francisco	I-580 WB	Bellam Boulevard	CA	0.38	65	8.1
3	New York	Cross Bronx Expressway/I-95 SB	I-895/Sheridan Expressway/Exit 4A	NY	0.55	93	11.9
4	New York	Cross Bronx Expressway/I-95 SB	White Plains Road/Exit 5	NY	0.27	87	12.3
5	New York	Harlem River Drive SB	3rd Avenue	NY	0.15	81	12.4
6	New York	Van Wyck Expressway/I-678 NB	Liberty Avenue/Exit 4	NY	0.58	77	13.1
7	Los Angeles	Hollywood Freeway/U.S. 101 SB	Vermont Avenue	CA	0.64	77	14.0
8	Chicago	Dan Ryan Expressway/ I-90/I-94 SB	Canalport Avenue/ Cermak Road/Exit 53	IL	0.52	77	13.6
9	New York	Harlem River Drive SB	2nd Avenue/125th Street/Exit 19	NY	0.23	84	12.5
10	Chicago	Eisenhower Expressway/I-290 EB	U.S. 12/U.S. 20/U.S. 45/Exit 17	IL	0.98	57	12.3
11	New York	Cross Bronx Expressway WB/ I-95 SB	Westchester Avenue/Exit 5	NY	1.15	77	14.5
12	Los Angeles	Hollywood Freeway/U.S. 101 NB	Los Angeles Street	CA	0.09	76	11.9
13	Los Angeles	Hollywood Freeway/U.S. 101 NB	Spring Street	CA	0.14	85	14.2
14	Los Angeles	Harbor Freeway/I-110 NB	Adams Boulevard	CA	0.13	73	15.8
15	New York	George Washington Bridge EB/ I-95 NB	Center Avenue	NJ	0.14	68	9.0
16	New York	I-95 NB	U.S. 1/U.S. 9/U.S. 46/Exit 72	NJ	0.42	66	9.7
17	New York	Harlem River Drive NB	Lower Level Washington Bridge	NY	0.09	74	10.3
18	Los Angeles	Hollywood Freeway/U.S. 101 NB	Alameda Street	CA	0.26	73	14.0
19	Chicago	Dan Ryan Expressway/ I-90/I-94 NB	Ruble Street/Exit 52B	IL	0.13	76	16.1
20	Los Angeles	Hollywood Freeway/U.S. 101 SB	Melrose Avenue	CA	0.31	68	15.9
21	Los Angeles	Harbor Freeway/I-110 NB	I-10/I-100/Santa Monica Freeway	CA	1.09	70	16.4

Table 1. (Continued)

2008 Rank	Area	Road/Direction	Segment/Interchange	State	Length	Hours Congested	Average Speed When Congested
22	New Haven	I-91 SB	I-95	CT	0.47	63	13.4
23	New York	Van Wyck Expressway/I-678 NB	Hillside Avenue/Exit 6	NY	0.27	79	14.4
24	New York	Van Wyck Expressway/I-678 NB	Atlantic Avenue/Exit 5	NY	0.47	75	12.7
25	Chicago	Dan Ryan Expressway/ I-90/I-94 NB	18th Street/Exit 52C	IL	0.34	75	15.7

Source: "*Inrix's National Traffic Scorecard: 2008 Annual Report,*" February 2009, http://scorecard.inrix.com/scorecard/request.asp.

More recently, an effort by a private data provider, Inrix, also identified the nation's worst bottlenecks (Table 1). Whereas previous bottleneck identification efforts were based on analytic procedures using traffic volumes and capacity data, Inrix's approach uses data assembled by them from a variety of sources. As direct travel time measurements become more common and better refined, the science of bottleneck identification and performance will improve.

Improvements Are Possible

"Seven of 18 bottlenecks identified in 1999 – including hotspots in Houston, Albuquerque, Denver, Boston, Los Angeles, and Washington, D.C. – no longer appeared on (subsequent) rankings of the country's worst chokepoints (due to) major reconstruction projects completed or underway."

American Highway Users Alliance, 2004.

WHAT IS FHWA DOING?

With regards to congestion, the Federal Highway Administration (FHWA) promotes a number of efforts to help reduce congestion on the nation's highways. Together with our state partners, who implement these strategies, these efforts can allow for more informed decisions, better coordination, and quicker actions to mitigate the problems.

Recurring Congestion Program Strategies

- **Tolling and Pricing.** Value pricing entails fees or tolls for road use which vary by level of vehicle demand on the facility. Fees are typically assessed electronically to eliminate delays associated with manual toll collection facilities.
- **Public-Private Partnerships.** "Public-private partnerships" (PPP) refer to contractual agreements between a public agency and private sector entity that allow for greater private sector participation in the delivery of transportation projects. FHWA is working with our

partners in the public and private sector to further investigate these promising partnerships.

- **Real-Time Traveler Information.** This is "decision-quality" information that travelers can access, understand, and act on to choose the most efficient mode and route to their final destination. Timely and detailed information about traffic incidents, the weather, construction activities, transit and special events, all aid in improving travel time predictability, better choices, and reduced congestion.
- **Traffic Signal Timing.** Signal timing should correspond to the current traffic patterns. Often signals are initially timed, but not re-adjusted when traffic patterns change. This results in inefficiency and unnecessary delays. Goal: work with state and local agencies in congested metropolitan areas and encourage best practices for improved traffic signal timing.

Nonrecurring Congestion Program Strategies

- **Traffic Incident Management.** This utilizes a combination of public safety functions and traffic management functions – it requires cooperation between various public agencies to reduce congestion by clearing traffic crashes and removing stalled vehicles. FHWA is championing laws, policies, and practices that speed up the clearance of major and minor incidents that create congestion.
- **Work Zone Mobility and Highways for LIFE.** The Work Zone Safety and Mobility Rule advocates stronger consideration and management of work zone safety and mobility impacts to reduce congestion during construction projects. The Highways for **Long**-lasting, **Innovative**, **Fast** construction of **Efficient** (also known as HfL) and safe pavements and bridges pilot program as outlined in SAFETEA-LU is meant to accelerate the rate of adoption of innovations and technologies, thereby improving safety and highway quality while reducing congestion caused by construction.

The Localized Bottleneck Reduction Program

The Localized Bottleneck Reduction (LBR) Program promotes operational and low-cost bottleneck mitigation strategies to improve mobility in the short term. Managed by the Office of Operations, the program serves to bring attention to the root causes, impacts, and potential solutions to traffic chokepoints that are recurring events; ones that are wholly the result of operational influences. The goal of the program is to raise awareness of bottlenecks at the state level and promote low-cost, quick-to-implement geometric and operational improvements to address recurring chokepoints. The LBR Program has several activities underway, including:

- This Primer, which overviews the wide range of operational and low-cost strategies available to reduce congestion at bottlenecks.
- A compendium of state best practices in bottleneck identification, assessment, countermeasures, and evaluation, including how bottlenecks are treated in the annual planning and programming processes.
- Regional workshops for state and local agencies to learn and share information on localized bottleneck reduction strategies and how they can be incorporated into state and local planning processes.
- Guidance documents are forthcoming that are aimed at agencies and personnel who have first responsibility to address bottleneck congestion locations.

More Information on the Localized Bottleneck Reduction Program

The FHWA Office of Operations Localized Bottleneck Reduction (LBR) Program brings into focus the root causes, impacts, and potential solutions to traffic chokepoints that can be mitigated through relatively low-cost improvements. The LBR Program has several activities underway, including:

- A Primer on localized bottleneck reduction;
- A compendium of state-initiated solutions to localized bottleneck problems; and
- Regional workshops for state and local agencies to learn and share information.

More information is available at the FHWA Bottlenecks web site: http://ops.fhwa.dot.gov/bn/index.htm.

In the longer term, a "toolbox" will be developed. The toolbox will be a one-stop source for guidance on low-cost operational and construction strategies for improving bottlenecks. FHWA will also issue guiding principles and concepts common to low-cost operational improvement programs in order to further state efforts in adopting a comprehensive approach to addressing traffic congestion at bottlenecks.

Benefits of Localized Bottleneck Improvements

The LBR Program focuses on operationally influenced bottlenecks – small, localized "hot spots" where the design of the roadway itself becomes the constricting factor in processing traffic demand, resulting in recurring delays of generally predictable times and durations. Mega-projects required to resolve major bottleneck problems and systemic congestion (e.g., entire corridor rebuilds, multimile lane additions, and systemwide improvements) are far and above the focus of this program area. Unfortunately, when weighed against these larger, more visible projects, localized bottleneck problems often receive lower priority for funding or are put off entirely until they can be implemented as part of the larger, all-encompassing project. However, in this day and age of fiscal constraints, with agencies facing over-escalating costs and increasingly limited right-of-way, it is evident that "business as usual" in resolving congestion problems no longer applies. Low-cost bottleneck mitigations have several advantages that can help agencies deal with these developments:

- **They address current problems and therefore have high visibility.** Agencies are under increasing pressure to do something immediately about congestion problems. Because low-cost bottleneck treatments are small in scale, they can be implemented quickly, so benefits start accruing immediately.
- **They are highly cost-effective and usually have positive safety impacts**. Low-cost bottleneck treatments could mitigate or reduce crashes within weaving and merging areas, thereby increasing the cost effectiveness relative to safety merits.
- **They will be required as transportation funding for mega-projects becomes more constrained.** Major reconstruction projects are often justified as the only valid solutions to relieve congestion at the worst bottleneck locations. However, the cost of executing such projects is

usually enormous. Low-cost bottleneck improvements provide an effective way to stretch scarce resources.

- **Lower cost means more locations can be addressed.** More spot solutions can be implemented throughout a region, addressing more corridors than just a few large projects.
- **They are less invasive on the physical and human environments.** The environmental footprint of low-cost bottleneck projects is very low, both in terms of disruptions during construction and final design.
- **They are not necessarily just short-term fixes.** For some low-cost treatments, congestion benefits will play out over many years, not just a few. In fact, when combined with other forms of treatment (e.g., demand management and operations), they may be part of a long-term solution for a problem location or corridor.
- **They may be considered part of major reconstruction projects to address current problems.** Some state DOTs have successfully incorporated low-cost bottleneck treatments within the context of larger, multiyear reconstruction projects.

FHWA Survey Suggests Opportunities for Lower Cost Solutions

Respondents to a FHWA Division Office survey reported that 71 percent of bottlenecks occurred on freeways, with 42 percent interchange related bottlenecks. Respondents cited that 36 percent of bottlenecks had no specific plans underway for improvement.

Full interchange or freeway reconstruction were the most commonly perceived solutions for resolving bottlenecks. However, at least a portion of these bottlenecks might be candidates for low-cost solutions that can be implemented in the short term to improve traffic flow.

Recent discussions with several state partners have reinforced their position. Specifically, high-ranking state DOT officials and mid-level staffers engaged in day-to-day operations have all opined that there are "tremendous" and "significant" benefits to pursuing low-cost operational improvements.

These benefits range from the direct (reductions in delay, increases in throughput) to the indirect (public confidence and agency image-boosting effects).

Identifying and Assessing Bottlenecks

Bottleneck locations can be identified through direct observation by agency personnel, aerial photographs, or video surveillance data. Some metropolitan areas have even conducted public outreach efforts to solicit input from motorists on bottleneck locations. Traffic analysis tools can systematically identify bottleneck locations by analyzing road segments for congestion or poor levels of service.

Once bottleneck locations have been identified, the root cause and severity must be determined. Special travel time runs, aerial photography, or video of suspected bottleneck areas can be used to pinpoint sources of operational deficiencies. On freeways equipped with detection technology, dynamic surveillance can be used to identify where and how often bottlenecks occur, and how severe they are. Archived traffic data can be used to measure whether the problem is growing or receding.

Sometimes, the operational cause of a bottleneck is evident, intuitive, or anecdotal. However, when multimile corridor congestion is prevalent, microsimulation modeling can assist in identifying, separating, and analyzing bottleneck dynamics within the corridor.

Data collection to support bottleneck analysis should be sufficient to capture the duration and extent of congestion. Typically, 15-minute traffic volume counts for all ramps and mainlanes for a four-hour peak period is adequate. Other data can be collected through travel time runs, video, or origin-destination studies.

Bottleneck analysis is necessary to study not only the subject location, but also the impacts of potential bottleneck remediation on upstream and downstream conditions. The analysis will justify action to correct bottlenecks, confirm the benefits of bottleneck remediation, or check for hidden bottlenecks along a corridor. When conducting bottleneck analysis, care should be taken to ensure that:

- Improving traffic flow at the bottleneck location doesn't just transfer the problem downstream. The existing bottleneck may be "metering" flow so that a downstream section currently functions acceptably, but the increased flow will cause it to become a new bottleneck.
- Future traffic projections and planned system improvements are inclusive in the analysis. Safety merits also should be strongly considered.

- "Hidden bottlenecks" are considered. Sometimes, the queue formed by a dominant bottleneck masks other problems upstream of it. Improving the dominant bottleneck may reveal these hidden locations. It is important to take into account the possibility of "hidden bottlenecks" during the analysis stage.
- Conditions not traditionally considered by models are accounted for. There are several bottleneck conditions, such as certain types of geometrics and abrupt changes in grade or curvature, that can't be analyzed by current analysis tools. Engineering judgment will need to be exercised to identify those problems and possible solutions.

Operation Bottleneck Solicits Public Feedback on Bottleneck Locations in Arkansas

Metroplan, the MPO for the Little Rock, Arkansas region, conducted Operation Bottleneck in October 2008, a public outreach effort designed to identify traffic bottlenecks as well as auto, bike, and pedestrian safety issues throughout the region. The program received 3,000 responses in 1S months, with on-line submissions constituting the highest return. Metroplan is currently reviewing, classifying, and analyzing the feedback, and the next step will be to coordinate with local jurisdictions.

How Do Bottlenecks Disperse?

Recurring congestion usually disappears when the crush volume drops back to a level that is manageable by the facility design. Nonrecurring congestion usually disappears when the dynamic event is removed. The rate at which merged lanes unite is strongly governed by the rate of dispersal at the constricted point (i.e., the "nozzle") and thus is less easily influenced on its own merits. In layman's terms, it hardly matters how well the two lanes "knit" together when there's no place to go! There is some evidence that in nonrecurring events, strong direction in the form of flagmen or specific messages (e.g., "Take Turns," "Merge Here," "Stay in Lane") may improve the rate of dispersal marginally. Such deployments are known as "Dynamic Lane Merges" (explained further below) as they rely on some aspect of proactive instruction. Realistically, the great majority of nonrecurring chokepoints are comparatively "static"; that is, there are signs forewarning the merge, but there is no proactive management of the merge. The information supplied at recurring chokepoints is even worse; barely much beyond a free for all. Finally, and to a lesser degree, the rate of arrival of vehicles upstream joining the back of the queue has some impact on the bottleneck.

"So You Think You Can Merge?"

Are you a "profiteering" lane merger, who seeks only your own personal gain to advance, or are you an "altruistic" driver who yields to others for the benefit of all?

Whether you are an "early" merger (upstream of the point of confluence) or "late" merger (at the last possible moment) does not brand you good or bad, Republican or Democrat, or even rightminded or left-minded. Anecdotally, both sides claim pre-eminence in their preference.

In the end, the prevailing conditions (i.e., whether the merge is executed at free-flow speeds or at a crawl; what guidance is provided) and the generalization that we are each left to our own devices as to consider when it is best to merge, conspire to make this more a study in human behavior, and less a study in efficiency.

UNDERSTANDING MERGING AT RECURRING BOTTLENECKS

The Difference in Merging for Recurring and Nonrecurring Conditions

Merging in a recurring backup is different than in a nonrecurring one. Outside of ramp metering, the former is essentially "cat herding" with implicit rules (often local in culture or habit) at best. Typically, not much guidance is given. Drivers just "suddenly" encounter taillights ahead, whereas, in a nonrecurring event, there is more apt to be instruction in the form of orange cones, signs, flagmen, or police. There is often direction to motorists *how* ("Take Turns"), *where* ("Merge Here"), and even *what* ("All Lanes Thru") to do/expect, and there can even be enforcement (of lane jumpers) or simply order (traffic cops) from chaos.

One might argue "What's the difference? I'm in bumper-to-bumper traffic regardless!" The great difference is the greater *potential* (in nonrecurring) for herding those cats. Nonrecurring congestion is better suited for empirical study, i.e., characterized by observation and trial instead of theory. The "Dynamic Lane Merge" (as we will see later) is a prime example of this. Conversely, recurring congestion is best suited for (and pursued by) the academic and scholarly; focusing on the root causes, the queuing theories, and the algorithmic "solutions" that might bend or break the cycle of recurrence. However, in real life recurring situations, drivers are mostly left to their own devices how and when to merge. Their *tendencies* may be studied, but "Exit 1/2-Mile Ahead" doesn't really help once you are in the backup. And therein lies the fundamental point; a recurring merge relies on a driver's predilection, as well as the aforementioned culture. For comparison, at nonrecurring events, an identical dynamic traffic control plan can be developed and implemented most anywhere and drivers will more or less treat it the same.

Which is Best? "Early" or "Late" Merging?

Can a better recurring merge be developed? Merging takes place at-speed or "at-crawl." The former is most often associated with free flow on-ramp maneuvers, while the latter is most often associated with bumper-to-bumper congestion. In either condition the motorist has the additional choice to merge "early" (upstream) or "late" (at point of confluence). This creates a matrix of four possible merge conditions; 1) at-speed "early;" 2) at-speed "late;" 3) at-crawl "early;" and 4) at-crawl "late." To further complicate things, guidance concerning where, when, and how best to merge can vary from modest to no forewarnings in *recurring* conditions to fully deployed Traffic Control Plans (TCPs) in *nonrecurring* conditions. Given that this primer is focused on the *recurring* bottleneck genre, the purpose of this section was to research if early or late merging was best for these noncontrolled situations; i.e., when no active TCP exists.

What Instruction is Given to Motorists?

Let's look at the difference between recurring and nonrecurring conditions. Work zones and special events are nonrecurring (random or irregular) conditions, wherein, dynamic TCPs are instituted for the expressed purpose of mitigating the planned event, and are almost universally mandated by the oversight agency. Orange cones, signs, flashers, and dynamic messages are typically deployed. The motorist is given forewarning and structured instruction how to negotiate the event zone. However, in recurring situations (chronic and habitual backups) there is minimal guidance. In the best case (e.g., a freeway lane drop) there are a couple of yellow standard warning signs and some pavement markings, which dutifully forewarn for free flow conditions, but are essentially superfluous in jammed conditions. At many other recurring congested locations (e.g., peak hour crowding onto ramps or through intersections, et al.) there is no guidance short of a free for all. This is because engineers build for, and sign for, the unconstrained condition. The result is that in many recurring situations drivers are rarely, if ever, given instruction *where* and *how* to merge, as they would be in a nonrecurring situation. This is important in framing the subject question; namely, in recurring, jammed conditions, is there academic evidence endorsing "early" or "late" merging?

Research could really only uncover data-driven studies in nonrecurring situations (wherein, motorists are more or less directed to follow signs in a work zone or event-driven scenario) and anecdotal discussions (in newspapers and the on-line community) in the recurring genre. Admittedly, in all of these cases, it mattered little in the discussion whether the cause was a recurring or nonrecurring event, but rather the fact that for whatever reason, jammed conditions existed and debate ensued. In the computer simulation community, driver behavior is a random variable used simply to enable the merge algorithms to execute and deduce rate of discharge, queue lengths and duration of the congestion. Again, the point being that in simulation, the queues merge – somehow – but no import is given *where* that merge occurs.

On the whole, drivers are typically left to their own strategies as to *how* to merge together at recurring chokepoints. Personal preference, impatience, frustration, speed-differentials, and other human and vehicular traits conspire to influence safety and reduce efficiency. Altruistic drivers are unselfish and yield – in varying degrees – to proactive drivers, who seek only their own benefit to cut in line. The only real conclusion that can be drawn is suggested by the similarity in methodologies used in the work zone studies. Specifically, in setting up "Dynamic Work Zones" (see "Studies: Post 2000" below) the mere fact that all of these trials presumed to set up – and study – a "late merge" scenario speaks to the engineering community's penchant towards this method over the "early merge" option for stop-and-go conditions. One theme, however, remained constant. Regardless of the amount of forewarning and direction given to motorists (e.g., "light" guidance in recurring situations and "heavy" guidance in nonrecurring situations) personal preference seemed to win the day. Absent absolute enforcement, motorists were observed to – or opined to – merge when and how they preferred, with less regard for any instruction. Anecdotal conversation and local opinion columns (e.g., "Traffic Doctor," "Mr. Gridlock," etc.) have occasionally debated the merits of merging early (upstream) versus late (at point of constriction). The fact that the debate even lingers gives strength to the notion that no consensus emerged from academic and Internet research. In Tom Vanderbilt's 2008 book "Traffic: Why We Drive the Way We Do (And What it Says About Us)" the author goes so far as to introduce the entire book with the "early" versus "late" debate, and tells why and how he became a convert to the late merge.

Early Attempts to Direct Motorists How to Merge

When the Interstates were built in the 1960s and 70s there was often "instruction" by local engineers and the media of how to engage Interstate ramps, acceleration and deceleration lanes, etc. Of course, at that time, traffic was less congested on the whole, and the merging and diverging were essentially lessons in how to enter and exit Interstates. Academia has touched on queue theory, gap analysis, and related safety-oriented aspects, but none of these studies have focused much on educating motorists how to merge efficiently, unless one considers a "queue" or a "traffic stream" as an entity that can deduce instruction. Nevertheless, the academic community has essentially confirmed, via queuing theory and micro simulation that the discharge rate after the merge governs congestion on the segment. In layman's terms, there is a finite capacity of the single lane downstream of the constriction. Very little of what happens upstream can refute the laws of physics; that only one vehicle can occupy the discharge space at a time; and in a jammed situation, the lead vehicle does so from essentially a crawl speed.

Studies: Post 2000

There have been studies and trials of Dynamic Lane Merge Systems (DLMS) or similar, wherein detectors and variable message boards are designed "to utilize the best aspects of the early and late merge strategies" (Minnesota DOT, 2004).

Briefly, the system switches "on" for late merges, and "off" for early merges. The Minnesota study, plus earlier, similar studies by Michigan,

Kansas, and Maryland, was conducted in nonrecurring situations; specifically, in work zones. Generally, once queues were detected as growing, a series of dynamic messages would kick in that would direct motorists where and how to merge during the stop-and-go durations. Typically the messages advise motorists to "Stay in Lane," "Merge Here" (at the end of the dropped lane), "Fill All Lanes," and even "Take Turns" or similar. As the queues clear, the messages would turn off, returning the decision of how to merge upstream of the work zone back to the drivers. Arguably, the real test factors driving these trials are safety and increasing throughput in work zones (i.e., in nonrecurring conditions); that is to say that none of the trials was necessarily testing motorists' preferences during the relatively more benign recurring condition.

The mere fact that in jammed conditions, all of these trials presumed to set up a "late merge" scenario speaks to the engineering community's penchant towards this method over the "early merge" option for stop-and-go conditions. The converse seems also to provide endorsement; namely, that absent jammed conditions the late merge is "relaxed" and early merging is "allowed." Note that we don't use the word "encouraged" for the latter condition; we merely note that the decision of how and when to merge is abdicated to the driver.

The 2004 Minnesota study/comparison of four different deployments (locations, conditions) yielded promising success only in terms of queue reduction and driver behavior. There is anecdotal evidence in the study to support that drivers are less aggressive, and perceive shorter travel times when the overall distance traveled under congested conditions is reduced. However, all things being equal, the rates of dissipation downstream of the constriction don't seem heavily influenced by "early" or "late" merging in stop-and-go conditions, as evidenced by the study's finding that 1,600 vehicles per hour per lane downstream of the constriction seems a maximum flow rate for any deployment – a number that not coincidently is the suggested maximum flow rate listed in the Highway Capacity Manual for "short-term freeway work zones." (The reader is cautioned that the conclusions herein are merely the observations of the Minnesota study authors and should not be considered tantamount to a broad brush conclusion for all bottleneck conditions and factors.) Further, these deployments, with their garish orange barrels, cones, and VMS boards, scream "temporary condition," and are expensive to deploy on an ad-hoc basis – not the kind of thing that one could expect to be adopted as standard practice everywhere at recurring chokepoints.

Excepting for some basic, generic instruction in states' drivers manuals ("wait for a safe gap in traffic" – typ.) little has been done at the national level to educate drivers how to merge safely and efficiently, as compared to other national education efforts promoting seat belt compliance, school zone safety, traveler information, or pedestrian rights and practices. The perceived reason for this may simply be the expectation that there will always be drivers who feel they know best how and when to merge in a queue, irrespective of any instruction to the contrary. The altruistic view is to leave gaps, yield to your neighbor, take your turn but don't force your turn, and generally don't deny him or her entry into your lane. The more proactive view is to take first opportunity to cut in line, perhaps "line jump" to chase whichever line seems to be moving, and scuttle the principles of any orderly manner. Anecdotal evidence from many local traffic blogs and an Internet search finds strong sentiment from both camps as to why they think their method is best.

Merging "At-Speed" versus Merging "At-Crawl"

Under free flow conditions, the Highway Capacity Manual states that 65 mph freeways are capable of enabling up to 2,300 pcphpl – a number that can be compared to the 1,600 pcphpl "work zone" capacity-restricted volume above. (The 2,300 number takes into account safe vehicle separations and optimum conditions.) This roughly 30 percent reduction in capacity is attributable to the fact that in stop-andgo conditions, the inefficiencies of merging are magnified.

Whether or not those two adjacent lines merge early, late, or otherwise, they are still metered by the fact that only one vehicle can proceed past the constriction at a time. As upstream volumes dissipate, free flow speeds increase, and vehicles enjoy a "head start" of momentum through the opening, allowing for slightly higher discharge rates that would fall above 1,600 pcphpl but cap at 2,300. In layman's terms, merging at-speed upstream of a constriction slightly improves the discharge capability, but there is still a finite time necessary for all vehicles to clear in one lane downstream.

Merge Principles

Can we increase the efficiency of merging prior to the discharge point? In two words – *be orderly*. Not surprisingly, safety improves too. It is repeatedly shown that traffic is inherently safer when all vehicles are traveling at or near the same speed.

Think of an orderly progression on a crowded escalator. Everyone is safely cocooned because they are going the same speed. Now imagine the bumping and chaos that would occur if impatient folks pushed past others.

For What it's Worth

Webster's on-line dictionary defines "merging" (using 'traffic' as the colloquial) as "to blend or come together without abrupt change." (Emphasis added)

Principle #1: "Go Slow to Go Fast"

"Go slow to go fast" is an increasingly trendy expression in traffic circles. It speaks to the seemingly paradoxical idea that if we slow down the rate of our "mixing" we can get past a constriction faster. A well known example (actually the winning entry in a 2006 contest to demonstrate the meaning of "throughput maximization") is the "rice" experiment. In the first case, dry rice is poured all at once into a funnel.

In the second case, the same amount is poured slowly. Repeated trials generally conclude about a one-third time savings to empty the funnel via the second method. And, it should be noted, there is a tipping point reached as one graduates from a v-e-r-y slow pour, to a medium pace, and so on. What lesson does the rice experiment teach us about traffic?

The densely packed rice (or traffic) in the first trial creates friction in the literal sense and the practical sense, respectively. The denser the traffic, the smaller the safety cushion around each driver, and the more cautious (i.e., slower) he becomes.

A classic "bell curve" diagram also serves to explain how traffic throughput reaches an apex up to the point where traffic friction and conflict conspire to begin a decline in the rate of throughput and speed. There exist some examples of validation of this principle at intersections (e.g., traffic signalization, roundabouts, vehicle detection) that demonstrates that slowing or stopping some traffic benefits the aggregate flow, and is far better than the free-for-all converse.

In the bottleneck and corridor genres, we have ramp metering and speed harmonization (see below), respectively, providing examples on freeways.

Principle #2: Keep Sufficient Gaps

Keeping sufficient (or ideally, the largest possible) gaps leads to uniform and free(er) traffic flow. Gaps allow for small adjustments in braking, accelerating, and drifting. The larger the gap, the lesser the "ripple" affecting adjacent and following vehicles, which otherwise would react by slowing. Gap maintenance is achieved by rote in high-occupancy vehicle (HOV) lanes or high-occupancy toll (HOT) lanes; by selective admittance in the former, and by dynamically shifting the price every few minutes in the latter. The target benefit is to allow qualifying vehicles the guarantee of a free flow trip, versus the hit-or-miss prospect in the adjacent general purpose (GP) lanes. Both cases have the added (and intended) benefit of removing vehicles and or person-trips from the GP lanes too; so all traffic streams win when these practices are employed. Absent out-and-out violators who can muck up the system, agencies can tweak the lane mandates to keep the systems running at optimum levels.

How does this apply to localized bottlenecks? Theoretically, the same "gapping" principles would hold true in backups; to wit, leaving progressively larger gaps would allow for progressively better progression. (Taken to the extreme, no "bottleneck" would even exist!) The point is that in congested situations the constant brake-tapping in bumper-to-bumper traffic works to self-perpetuate the problem. No one can get much momentum before he or she has to react to the vehicle directly ahead or adjacent. The ripple effects are short, abrupt, and inefficient. The obvious problem with this is that human nature simply won't allow for the patience and orderliness to make this work. The second I create a sufficient gap between me and the car ahead, some "profiteering" lane jumper will fill it. Which is a nice segue into the next principle; zippering.

Principle #3: Zippering

Unlike principle #2, which is noted to be fairly impractical to expect, this one could easily be melded into our regular practice; namely, to take turns, or "zipper" merge at the front of the line. The fairness – and simple visualization – of this principle speaks for itself. And there is already precedence that we have been schooled in this; witness the "Yield" condition and many recurring locations where this is the unwritten rule; newcomers quickly adapt! Advocates of zipper merging are proponents of "late" merges; i.e., staying in your lane until the last possible moment and taking turns to get through the chokepoint nozzle. One enterprising fellow in California has gone so far as to adorn his car with a zipper graphic and messages promoting this method.

But is this latter day Don Quixote merely tilting at this traffic “windmill”? To be fair, “late” merging seems the slightly favored method in web blog research, although as noted prior, there are unyielding stout proponents of the early merge too. Let’s take a closer look.

Imagine two lanes of cars. The left lane (L) is the continuous lane and the right lane (R) is the dropped lane. You are 6th in line in the R lane. If everyone stays put, and the zipper order is L, R, L, R. etc., your neighbor to the left will be 11th and you will be 12th to merge. If, however, you “early merge” and cut in front of him into the L line (and your fellow motorists in the R line all move up in succession) then you will become 11th, the person originally behind you (formerly 14th) now becomes 12th, and your neighbor will have dropped to 13th. You win. Your neighbor loses. But the real winner is the guy originally behind you; he’s gained two spots.

Now consider the same scenario except the zipper order is R, L, R, L... etc. In the orderly scenario you would be 11th and your neighbor would be 12th to merge. If you cut in front of him as in the first scenario, the guy behind you (who moves up to fill your space) becomes 11, you are 12, and your neighbor is slotted at 14. You lose. Your neighbor really loses (drops two slots) and the guy behind you (formerly 13) really wins; he gains two spots – again.

Congratulations! In both scenarios, by “early merging” you have definitely improved the slot for the guy behind you! You may or may not have improved your slot. And either way you made your neighbor mad! Of course, in real life, the permutations are mind boggling when one considers multiple jockeying for position in a long queue by you and others, or the dynamic created by multiple lanes. You may think you are making progress one line jump at a time, but your move(s) may have been canceled by someone way up ahead of you; or the woman opposite him. And so on. It effectively becomes akin to playing the lottery – you are at the mercy of others’ actions equal to, or more so, than your own.

And all for the perceived benefit of getting a few spaces ahead through the nozzle. Of course, there’s another way to look at it too; it’s all a wash when said and done. It’s hard to know. All you may have accomplished by line jumping is to elevate yours or others’ stress. Better to sit, wait, take your turn, and put your radio on some calming music.

Principles Put into Practice: Variable Speed Limits and Speed Harmonization

Variable speed limits (tried in work zones; i.e., nonrecurring conditions) and the European concept of "speed harmonization" (nonwork zones) both intend to "harmonize" traffic by regulating speeds. In the latter case, a series of overhead gantries gradually adjust speeds through congested highway segments in order to flatten the sinusoidal effect of traffic speeds bouncing between open sections and interchanges. Speed harmonization is typically effected as the open highway approaches the denser central business district. A great expense is incurred by the cost of the overhead, spanned gantries, the necessary detectors, the interconnectivity, the necessary operational overhead, and the sheer number of gantries required along the multi-kilometer corridor; a smart highway, as it were. At the end of the day however, such a system is not transferable from facility to facility, so you had better have good justification and need to install it where you did. But that's the bad news too; that such pervasive congestion exists in the first place. What if the congestion disappears (e.g., a parallel facility opens up or a large traffic generator goes away)? The agency is stuck with an expensive boondoggle.

Specific to the question at hand, speed harmonization does not seek to control driver behavior to switch lanes, or mandate how to merge, but rather only to control speeds. "Go slow" (harmonize) as a means to move more traffic than otherwise might have gotten by. As of 2009, this system has not been employed in the United States, although there is intent to try it in the Seattle, Washington area as part of a Federally funded demonstration project.

So How Does One Merge Safely?

What methods work best for motorists to merge traffic efficiently in bottlenecked conditions? The answer, it seems, is largely – perhaps only – dependent on the flow rate of the continuing lane and less-so the dropped lane next to it. The debate between "early" or "late" merge seems more appropriate for human behavioral studies of aggressiveness, impatience, self-indulgence, and simple preference for one method over the other. However, there is general agreement in the engineering community on which method works best for each of two scenarios. Early merges work best under "Scenario 1" (see following page) when traffic is generally free-flow. Late merges seem to work best under "Scenario 2" when traffic is stop-and-go.

Scenario 1: In at-speed, or nearly free-flow conditions, it is best to merge upstream of the constriction

Motorists in the thru-lane: Allow gaps. By allowing gaps you allow the merging vehicle the opportunity to adjust his/her speed to fill a gap.

Motorists in the dropped lane: Keep up speed and fill gaps at-speed. Use the dropped lane as an acceleration lane; i.e., as an opportunity to match speed with upstream traffic and merge at-speed at earliest and best opportunity.

Benefits: Traffic is safest when all vehicles travel at or near the same speed.

Potential pitfalls: 1) When a vehicle in the dropped lane can't or won't move into the through lane in time, then the dropped lane ends, forcing the vehicle to suddenly slow or stop. This will invariably cause a dangerous ripple as the stopped vehicle now must join moving traffic from nearly a dead stop; at risk of causing a through vehicle to slow to allow it in. 2) Pitfall #1 above also has the potential to cause following vehicles to queue behind the "stopped" vehicle due to an increasingly shortening merge lane. Motorists now forced to slow in the dropped lane become anxious to merge and may cause further ripples. If not allowed to remediate itself, pitfall #2 will morph into "Scenario 2" below. 3) Motorists in the thru lane won't yield (usually by way of speeding up).

Bottom line: Relax. Allow motorists to merge at-speed or fill gaps as opportunity allows.

Scenario 2: In heavy congestion (stop-and-go conditions) it is more efficient to fill both lanes and zipper-merge at the point of constriction

Benefits: Drivers experience less stress when they understand that each in-turn will have a chance to get past the point of constriction. The zipper merge is orderly and fair.

Potential pitfalls: 1) Queue jumping. This can occur in a number of ways; a) by bypassing cars in line, the "jumper" infuriates others because he is out of turn; or b) by having a second or third vehicle try to fill a single gap, this frustrates altruistic motorists who now feel they are being taken advantage of. 2) Lane blocking.

This occurs when a motorist in the continuous lane fails to fairly permit his neighbor to zipper ahead. 3) Line jumping back and forth, as it recalls the old axiom "the other line is always moving fastest."

This only serves to disrupt the continuity of the merge ahead. Bottom line: Remain orderly and fair. Do not line jump.

LOW-COST BOTTLENECK IMPROVEMENT STRATEGIES

Here is a sampling of short-term, low-cost operational and geometric improvements. All of these remedies address operational deficiencies, as opposed to other congestion mitigation efforts that address driver choice, travel demand, corridor-wide upgrades, or simply (but expensively) building our way out of congestion.

The bottleneck solution, it seems, does not lie with the mechanics of merging, but rather in reducing the underlying impacts in the first place.

1) **Shoulder conversions.** This involves using a short section of traffic bearing shoulder as an additional travel lane. Shoulder conversions are appropriate between interchanges or to provide lane congruency with adjacent sections. The shoulder condition be rated for use as a travel lane.
2) **Re-striping** merge or diverge areas to provide additional lanes, provide an acceleration/deceleration lane, extend the merge/diverge area, or improve geometrics to better serve demand.
3) **Lane width reductions.** This involves reducing lane widths and re-striping to add an additional travel and/or auxiliary lane.
4) **Modify weaving areas** by adding collector/distributor or through lanes.
5) **Ramp modifications.** These could include ramp metering; widening, extending, closing, or consolidating ramps; or reversing entrance and exit ramps to improve operations.
6) **Speed harmonization (variable speed limits).** This is the practice of adjusting speed limits when congestion thresholds have been exceeded and congestion and queue forming is imminent. Speed harmonization can also be used to promote safer driving during inclement weather conditions. This mostly European practice reduces the traffic "shock wave" that results through congested corridors, thereby delaying the onset of a breakdown in traffic conditions.

The result is decreased headways and more uniform driver behavior, which indirectly benefit bottlenecks and chokepoints.

7) **"Zippering"** or self-metering that promotes fair and smooth merges. A motorist who is 10th in line knows that he will be 20th to merge into the single lane ahead. This helps to eliminate line jumpers that bull ahead, disrupt the queues, and often block adjacent lanes until they force their way in line. Usually this method of merging requires on-site enforcement, but often is exhibited by regulars who know the process and are willing to abide.
8) **Improve traffic signal timing** on arterials. Also, traffic signal timing improvements at ramp terminal intersections will prevent ramp queues from backing up onto freeway mainlanes.
9) **Access management principles** to reduce vehicular conflicts (hence, delays) on arterial corridors
10) **Continuous flow intersections.** These are unconventional at-grade intersections which eliminate one or more left turn conflicts at a main intersection. This is achieved through dedicated left turn bays located several hundred feet prior to the main intersection which allow left turning vehicles to move at the same time as through traffic. The left turn traffic signal phase is eliminated, allowing more vehicles to move through the main intersection and thus reducing traffic congestion and delays. These at-grade intersections achieve traffic flow similar to grade-separated interchanges, but at a considerably lower cost.
11) **High-Occupancy Vehicle (HOV) or reversible lanes.**
12) **Provide traveler information** on traffic diversions.
13) **Implement congestion pricing.** Congestion pricing entails charging fees or tolls for road use that vary by level of vehicle demand on the facility. The objective is to bring supply and demand into alignment. As public acceptance grows and legislative restrictions are relaxed, congestion pricing will increasingly be viewed by transportation practitioners as a powerful and relatively easy to implement strategy to address bottleneck congestion.

In 2006, as part of the research conducted for National Cooperative Highway Research Program Project 3-83 ("Low-Cost Improvements for Recurring Freeway Bottlenecks"), a series of interviews was conducted with state and local transportation agencies to assess the effectiveness of low-cost improvements used at bottleneck locations within their jurisdictions. Table 2 was developed from these responses.

Table 2. Evaluating Bottleneck Improvement Effectiveness

Bottleneck Types	Mitigation Measures									
	Right Shoulder Conversion	Left Shoulder Conversion	Lane Width Reduction	Auxillary Lanes	Collector-Distributor Road	Re-Stripping to Add More Narrow LanesAll Purpose Lane (Concurrent or Reversible)	HOV Lanes (Concurrent or Reversible)	Ramp Metering	Temporary Ramp Closures	Traveler Information
Heavy On-Ramp Demand	◒	○	◒	●	○	●	○	●	○	◒
Weaving Sections	◒	◒	●	◒	◒	●	●	○	○	◒
Lane Drops	●	○	●	●	○	●	●	◒	◒	◒
Tunnels and Bridges	○	○	●	○	○	○	◒	○	○	◒
Horizontal and Vertical Curves	●	●	○	●	○	●	◒	◒	◒	◒
Narrow Lanes and Lateral Obstruction	◒	◒	○	◒	◒	◒	◒	◒	◒	◒
Inadequate Accelerated and/or Decelerated Lanes	●	○	◒	●	●	◒	◒	◒	●	●

● = good solution ◒ = may be helpful ○ = not applicable

The results showed that agencies are using a wide range of strategies to improve bottlenecks, many of them low-cost improvements that can be implemented quickly. The most frequently used operational improvements were ramp metering, auxiliary lanes, and HOV lanes.

Some of the key questions and considerations when selecting improvement alternatives for bottleneck removal include:

- Is there an inside shoulder that would create a usable traffic lane for a short section of freeway?
- If a shoulder is considered for conversion, is there right-of-way (ROW) to allow adding one back for part of the length of the project?
- If there are bridges, are they wide enough to accommodate the extra lane while allowing adequate clearance to barriers (2 feet) and an outside shoulder? If not, are they short enough that a loss of shoulder as a breakdown lane would not be critical (500 feet or less)?
- If changes to an entrance or exit ramp or weaving area are considered, will adjusting the position of ramp gores cause geometric problems which must be resolved?
- Are vertical clearance issues, grade-matching, and sight distance problems created?
- If the bottleneck movement itself cannot be fixed reasonably, can the other traffic which is affected by it be better accommodated?
- Finally, will the improvement invite enough new traffic to cause immediate breakdown again or is this truly the clearing up of a "kink" in the system, without being a capacity addition which will overload some other part of the facility?[1]

Evaluating Bottleneck Improvement Effectiveness

After implementation, it is often beneficial to conduct an "after" evaluation to gauge the effectiveness of the bottleneck removal project. A conservative approach to evaluating treatment effectiveness is to evaluate operational and safety benefits achieved. Common evaluation methods include microsimulation modeling, cost/benefit analysis, and crash data analysis using

[1] These options quoted directly from recent work by the Texas Transportation Institute: Freeway Bottleneck Analysis Methodology.

data collected before and after project implementation. In addition, the following performance measures are often used to assess the effectiveness of bottleneck improvement strategies: average speed (travel time), lane density, queue lengths, queue discharge rates, vehicle miles of travel (VMT), and vehicle hours of travel (delay). Additional insight could be obtained from before and after opinion surveys of area drivers. These types of evaluations are often not done, yet are important to quantify the benefits achieved through bottleneck mitigation.

POTENTIAL ISSUES WITH BOTTLENECK TREATMENTS

Because some bottleneck treatments involve innovative solutions that maximize effectiveness with a minimum of new construction, they are occasionally at odds with highway design standards. For example, the addition of slip ramp to a collector/distributor road or the use of a shoulder a through lane at selected locations may not strictly follow allowable design standards. Such deviations have the potential to degrade safety if not properly implemented; the elimination of a shoulder may led to more collisions with roadside features or may impede incident management activities. As it is FHWA's intent to foster creative approaches for low-cost bottleneck improvements, agencies should not see the design standard issue as insurmountable.

Rather, they should fully assess the potential safety impacts of strategies and devise ways of addressing them, if necessary. For example, in the case of a shoulder-to-lane conversion, review of crash data, and the specific roadway location (perhaps through a Roadway Safety Audit), it may be determined that a barrier is required to keep vehicles off of the roadside. It may also require a change in incident management policy that would allow emergency vehicles to access incidents from the opposite direction. Finally agencies should be in contact with the FHWA Division offices throughout the process as design review may be required, depending on circumstances.

The second potential issue relates to air quality conformity. Because they are short-term in nature, localized bottleneck improvements may emerge as formal projects that have not been previously identified in Statewide Transportation Improvement Programs or MPO-generated Transportation Improvement Programs. Thus, they may not be part of those projects that have been approved to deal with air quality issues in the region or state. Such occurrences must be dealt with on a case-by-case basis by agencies wishing to undertake bottleneck improvements. One point worth noting: if air quality conformity in a location precludes or discourages major capital expansion (e.g., additional lane-miles), the type of improvements in a localized bottleneck program clearly do not fall in this category.

EXAMPLES OF HOW AGENCIES ARE DEALING WITH BOTTLENECKS

Many transportation agencies have recognized that low-cost treatments can provide effective congestion relief at bottlenecks. A wide variety of improvements have been implemented and many innovative improvements are emerging. This following section provides expanded explanations of how these transportation agencies used strategies to improve congestion at bottlenecks.

- Low cost bottleneck improvements in **Minnesota**;
- **Connecticut DOT's** low-cost efforts to improve a freeway diverge area;
- Expediting bottleneck improvements ahead of major reconstruction project in **New Hampshire**;
- Continuous flow intersection improves intersection operations in **Utah**; and
- Operation Bottleneck Relief in **Los Angeles, California**.

For a more complete listing of representative projects, or for additional background on specific projects, please visit the FHWA "Bottleneck" web site at http://www.ops.fhwa.dot.gov/bn/index.htm or contact the Localized Bottleneck Reduction Program manager.

Low Cost improvements in Minnesota Provides Similar Benefits as a Mega Project, but at a Vastly Different Cost

In 2006, Minnesota DOT utilized low-cost strategies to resolve a bottleneck on State Highway 100 between 36th Street and I-394 in St. Louis Park, a suburb west of Minneapolis. This section of highway had only two lanes, while upstream and downstream sections of SH 100 had three lanes.

A prior study found that SH 100 in this area was congested for six hours out of every day, the longest duration of congestion than on any other freeway in the Twin Cities metropolitan area.

A full third lane was added in each direction by narrowing the left and right shoulders. In addition, a nearby diamond interchange was connected via collector-distributor roads, reducing access points on SH 100 from seven to four.

The improvements eliminated weaving movements that existed along this stretch of on- and off-ramps past the city.

The result yielded a (some say conservative!) 13:1 benefit/cost ratio; an increased throughput of 14,400 vehicles during peak periods daily; and a reduction in backups from 5 to 6 miles previously (depending on direction) to one-quarter mile. Minnesota DOT received an outpouring of positive public reaction, including one local newspaper's "Public Project of the Year" award.

Remarkably, this $7.1M project accomplished the same results as a $138 million project on a parallel freeway facility, IH-494 from Highway 5 to Highway 55.

Granted, the IH-494 design/build project was closer to a total facility rehabilitation with noise walls and other costs, but the effective elimination of long bottleneck queues was realized with the SH 100 project for a fraction of the cost.

Re-striping Resolves Bottleneck Conditions at Freeway Diverge Area in Connecticut

Congestion routinely existed on I-84 eastbound in Danbury in advance of the I-84/Route 7 split. The existing lane configuration in the vicinity of the exit ramp consisted of the left lane as an exit-only lane, with the center and right lanes continuing as through lanes.

Because significant traffic exited onto Route 7 northbound, this arrangement created weaving problems and bottlenecking, as all traffic bound for Route 7 had to maneuver into the left lane to access the exit.

Connecticut DOT devised a low cost solution which involved re-striping the lanes within the diverge area to better serve demand.

The I-84 eastbound lanes were reconfigured to allow the center lane to function as an option lane, where vehicles would have the option to exit or continue on I-84.

It was determined that the existing pavement width in the vicinity of the ramp gore would accommodate the option lane and would tie in with the current lane configuration on the ramp, which included two lanes.

The project required minor signing and striping revisions and some pavement modification. The bottleneck improvements will be completed as part of a larger interchange improvement project at I-84/Route 37, which will be advertised in June 2009.

The cost of the bottleneck improvements is estimated to be $900,000.

I-84 Eastbound at Exit 7 (Route 7 Northbound) – Existing Lane Configuration

I-84 Eastbound at Exit 7 (Route 7 Northbound) – Proposed Lane Configuration

Bottleneck Improvements Expedited Ahead of Major Reconstruction Project in New Hampshire

The existing 19.8-mile four-lane section of I-93 from the Massachusetts state line north to Manchester was originally built in the 1960s and since that time, the facility has not undergone any major system wide upgrades or modernization. The interchanges at Exits 2, 3, and 5 remain essentially as they were originally constructed. Exits 1 and 4 were partially reconstructed in the late 80s and early 90s, but are rapidly becoming incapable of meeting traffic demand.

To address these problems, NHDOT identified a major, multiyear reconstruction project.

The project will include: widening the mainline from two to four lanes per direction; widening of 44 bridges; reconstruction of all five interchanges on the section; addition of sound barriers; construction of three new park and ride lots; expanded bus service; and the purchase of right-of-way for a future light rail line. However, because this project has an ambitious scale, it will take several years to complete and major congestion now occurs at Exit 5.

The State of New Hampshire has selected work at Exit 5 (Londonderry) on I-93 to be expedited in order to address the bottlenecking concerns which currently exist. The signalized northbound and southbound ramps back up onto the mainline during peaks. The original project called for much of Exit 5 to be reconstructed in FY 2009. In partial response to the LBR program attention, improvements to two of the existing ramps are being expedited. The southbound off ramp at Exit 5 will be extended. The shoulder area on the mainline will be widened and reinforced to hold some of the queue. The signal at the top of the ramp will be coordinated with loop detection. Reconstruction or widening of two nearby bridges will be accomplished. The estimated cost of this expedited work is $9.13 million. The completed project will meld into the overall 19.8-mile I-93 project

Continuous Flow Design Improves Intersection Operations in Utah

A study of SR-171 (3500 South) at SR-154 (Bangerter Highway) in West Valley City showed that the intersection is one of the busiest in the State, with over 100,000 vehicles passing through the intersection on a typical weekday. Heavy rush hour traffic forces drivers to wait for up to four cycles of the traffic signal to get through the intersection, which contributes to congestion and safety problems that ripple out to affect other intersection on both 3500 South and Bangerter Highway. The study estimates that traffic through the intersection will increase 31 percent by 2030.

Several improvement alternatives for the intersection were evaluated, including a grade separated interchange. However, UDOT decided to implement Utah's first continuous flow intersection (CFI). An unconventional intersection design, CFIs reduce the steps in the traffic signal cycle and place left turning vehicles along a safer path.

Compared to a grade separated interchange, the CFI is a low cost alternative.

The CFI improvement project was started in March 2007 and was completed six months later in September 2007, with a total project cost of $7.5 million.

Operation Bottleneck Relief in Los Angeles, California

The City of Los Angeles recently initiated this four-year program to identify and resolve bottlenecks at signalized intersections.

The City is severely constrained in expanding mainlines and adding additional lanes at intersections, as right-of-way costs are prohibitive.

The program has identified 300 bottleneck locations to date.

Bottlenecks are defined as signalized intersections reaching greater than 40 percent loop occupancy on the upstream system detectors for four 15-minute periods.

Once a signal is identified as a bottleneck, the signal timing and phasing is reviewed to see if these can be adjusted to meet demand better.

If this is not fruitful, then additional improvement types such as re-striping to provide additional lanes, parking restrictions, or physical widening are evaluated:

The program initially identified 98 intersections that met the occupancy thresholds during peak periods:

- Thirty six percent were successfully treated through signal retiming;
- Fourteen percent were identified as candidates for parking restrictions or lane additions;
- Eight percent were caused by ramp metering overflow from freeways, which will require coordination with Caltrans to adjust metering rates and/or provide extra queue storage; and
- Forty two percent could not be reasonably treated by any of the strategies identified above; demand was simply too high and improvements would be too costly.

Unfortunately, adaptive signal timing will not resolve the issues for the remaining bottlenecks on the arterial system, as the competing approaches and saturation requires alternatives beyond adaptive timing.

However, timing is always the first approach to resolution of the bottlenecks identified through the program.

The City anticipates adding 1,000 more signals over the next 15 years with 300 to 400 in the more immediate future.

The OBR program will be used in the long-range planning process to help identify funding sources and solutions.

Additionally, it will help building the case for changes both politically and within the community about how bottlenecks are addressed and resolved.

PLANNING AND PROGRAMMING BOTTLENECK IMPROVEMENTS

Unless transportation agencies make low-cost bottleneck improvements an explicit presence, it is likely that they will be overlooked or delayed; either deemed part of a "larger problem," or unnecessarily postponed to some indefinite out year. There are many ways to combat this:

- **Create a unique bottleneck program area.** By developing an annual "named" program, agencies can effectively identify, fund, and most importantly, advance low-cost treatments. A stand-alone program also has the added benefit of demonstrating to the public that the agency is actively engaged in fighting congestion.
- **Undertake occasional "special projects" to focus on bottlenecks.** Low-cost bottlenecks can be addressed through occasional "special projects." For example, the Minnesota DOT (MnDOT) conducted a "one-time" special compilation of projects meeting certain candidacy requirements. In much less than one year, MnDOT developed a highly accelerated process for bottleneck identification and prioritization, which led to many effective projects that were implemented in the following two years.
- **Integrate consideration of low-cost bottlenecks into existing programs.** Low-cost bottlenecks can be addressed programmatically even without a special program. By making them part of ongoing planning and processes, the can be part of an agency's congestion arsenal. For example, the Ohio DOT (ODOT) added a congestion-based index ranking to their predominantly safety-based Highway Safety Program (HSP). The result is that congestion weighted

problems now have a specific "voice" in the rankings of all candidate ODOT projects. The HSP was used as the mechanism because many of its projects also are low-cost and quickly implementable.

The following provide comparisons of how different state agencies have incorporated low cost bottleneck projects into their planning and programming processes:

- The **California Department of Transportation** (Caltrans) does not have a formal bottleneck planning process; rather, bottleneck issues are addressed at the district level as part of the regional planning process. Much of Caltrans' operational planning is guided by the Transportation Management System Master Plan, which sets forth the types of strategies that should be pursued in improving congestion. In much of California's metropolitan areas, traffic congestion is a 24/7 occurrence, and traffic management is a 24/7 job. Bottlenecks are tweaked "in real time" as part of their Corridor System Management Plans (CSMP), which are developed for some of California's most congested transportation corridors. System monitoring and evaluation is seen as the foundation for the entire process because it can not only identify congestion problems, but also be used to evaluate and prioritize competing investments. Caltrans does not have a direct funding for bottlenecks, although bottleneck projects are routinely programmed through the CSMP process.
- In **Ohio**, bottlenecks are part and parcel of the overarching Ohio Department of Transportation (ODOT) Highway Safety Program (HSP), which ranks all candidate projects and drives the statewide highway project selection and scheduling process. Beginning in 2002, ODOT developed a "congestion mapping" division that uses V/C ratios developed from traffic data recorders and roadway inventory. About the same time, ODOT administration pushed for an annual process of overlaying congestion-index and safety-index "hot spots." As a result, congestion hot spots now have a "voice" in the process regardless of crash indices, and congestion related problems now compete for attention in the HSP listing. Specifically, highway sections with V/C ratios greater than 1.0 are considered "congested" and are added to the listing. Sections with V/C between 0.9 and 1.0, but outside the cities of Columbus, Cincinnati, and Cleveland, are also added. After ODOT headquarters completes their statewide effort of

congestion mapping and safety indexing, the respective District engineers are responsible for developing countermeasures for their top-listed candidate projects. District Safety Review Teams sort projects into three scales – low (less than $100K and quickly implementable), medium ($100K to $5M and 1 to 2 years), and high (greater than $5M and necessitating more than years to implement) – and then compete with other projects having the same scale but in other districts.

- **Minnesota DOT** was originally driven to explore low-cost congestion relief projects because of budgetary restrictions, but soon realized that these projects could be implemented very quickly and, as a bonus, were highly visible and popular with the public. In much less than one year, MnDOT developed a highly accelerated process for bottleneck identification and prioritization, which led to many effective projects in the following two years. MnDOT also found that because of lower costs, it could identify multiple locations throughout the region and "spread around" bottleneck reduction projects in a fair and equitable manner. This process consisted of completing a study, which included a five-step process to narrow potential projects into a recommendation list to the state legislature. Evaluation of completed projects produced high benefit/cost ratios, usually greater than 8:1.
- The **Maryland State Highway Administration** (SHA) has a dedicated program of about $5 million per year for the identification and implementation of low cost traffic congestion improvements at intersections. The program's genesis tracks to when SHA asked "what can be done if and when a mega-project's 'no build' alternative is chosen?" The program has been well-received by the public and local governments. Projects typically include low-cost projects that can be implemented quickly, such as signal timing upgrades and adding turn lanes and through lanes at intersections. The Maryland SHA has also has had considerable success with projects to improve freeway ramps and merge areas that have reduced congestion bottlenecks at a low cost.
- In **Florida**, there is not a "bottleneck" planning process, per se; rather, bottleneck-related issues are addressed as part of the Florida Department of Transportation's (FDOT) standard planning process. The planning process, which is managed by the FDOT Systems Planning Office, begins with needs identification conducted at the district level, then projects are developed and proposed for the Cost

Feasible Plan. The Cost Feasible Plan is adopted and projects are ranked for inclusion into the 5-year or 10-year programs. Traffic data and the statewide model are used to identify deficiencies, but it is the responsibility of the districts to identify and resolve hot spots.

- **Washington State DOT** (WSDOT) has no direct funding for bottlenecks, but formally recognizes "bottlenecks and chokepoints" in their project planning and development process and devotes a portion of the Washington Transportation Plan (WTP) to them. At the planning stage, WSDOT considers bottlenecks together with traditional corridor improvements in a category called "Congestion Relief" – bottlenecks do not have their own category for assessment or funding. The Congestion Relief projects are ranked (prioritized) using the benefit/cost ratio and other qualitative factors.

Minnesota's Process to Identify and Prioritize Bottleneck Improvements

Step 1: Project Identification

Potential congestion management projects were identified from existing sources:

- Low-cost capacity improvements (e.g., auxiliary lanes)
- Re-striping lane configuration
- Traffic control device improvements (e.g., ramp meters and signal timing)

Step 2: Quantitative Screening

- Project cost < $15 million
- Not in three-year TIP
- Annual hours of delay > 25,000
- Minimum of two hours of congestion

Step 3: Qualitative Screening

- Design readiness
- Cost range
- Congestion benefit
- Construction traffic management
- Future demand changes
- No adverse downstream effects

Step 4: Expert Workshop

Projects were prioritized by an expert group during a half-day workshop.

Step 5: Project Planning

The following were prepared for each project:

- Geometric sketches
- Project scope
- Congestion impacts
- Safety impacts
- Benefit-to-cost ratio

WANT MORE INFORMATION?

The LBR Program is just one of several program areas dealing with congestion problems. More information may be found at FHWA's "Focus on Congestion" web page at: http://www.fhwa.dot.gov/congestion/links.htm.

DEFINITIONS

Auxiliary lanes – Typically, any lane whose primary function is not simply to carry through traffic. This can range from turn lanes, ramps, and other single purpose lanes, or it can be broadened to imply that a traffic bearing shoulder can be opened in peak periods to help alleviate a bottleneck, and then "shut back off" when the peak is over.

Bottleneck – There can be many definitions. Here are a few that are typically used.

1) A critical point of traffic congestion evidenced by queues upstream and free flowing traffic downstream;

2) A location on a highway where there is loss of physical capacity, surges in demand (traffic volumes), or both;

3) A point where traffic demand exceeds the normal capacity; and

4) A location where demand for usage of a highway section periodically exceeds the section's physical ability to handle it, and is independent of traffic-disrupting events that can occur on the roadway.

Capacity – The maximum amount of traffic capable of being handled by a given highway section. Traffic engineers usually speak in terms of "free flow" capacity.

Congestion (specifically, traffic congestion) – FHWA's Traffic Congestion and Reliability Report defines congestion as "an excess of vehicles on a portion of roadway at a particular time resulting in speeds that are slower – sometimes much slower – than normal or free flow speeds. (Congestion is) stop-and-go traffic. Previous work has shown that congestion is the result of seven root causes,[2] often interacting with one another." Since a bottleneck is a cause of congestion, congestion cannot be solely analogous to a bottleneck. Congestion is more. For example, a "congested" corridor may harbor multiple bottlenecks or any combination of the seven root causes of congestion.

Downstream traffic – Traffic that is beyond (past) the subject point on a highway.

Hidden bottleneck – A highway location where some type of physical restriction is present, but traffic flow into this area is metered by an upstream bottleneck so the location does not appear as a bottleneck under prevailing conditions. Removal of the upstream bottleneck will cause the hidden one to emerge as a new bottleneck.

Nonrecurring events – As it pertains to traffic, a delay caused by an unforeseen event; usually a traffic incident, the weather, a vehicle breakdown, a work zone, or other atypical event. Even if planned in many cases, like work zones and special events, they are irregular and not predictably habitual in location and duration.

Ramp metering – The practice of managing access to a highway via use of control devices such as traffic signals, signing, and gates to regulate the number of vehicles entering or leaving the freeway, in order to achieve operational objectives. The intent of ramp metering is to smooth the rate at which entering vehicles will compete with through vehicles. Done properly, ramp metering will calm the "mix" that occurs at these junctions.

Recurring event – As it pertains to traffic, a recurring event is a traffic condition (i.e., a bottleneck or backup) that one can presume to occur in the same location and at the same time daily, albeit for weekday or weekend

---

[2] The seven root causes are physical bottlenecks (a.k.a. "capacity constraints"), traffic incidents, work zones, weather, poorly timed signals et al., special events, and over-capacity demand (i.e., daily and seasonal peaks superimposed on a system with a fixed capacity). Some sources cite only six root causes because they see over-demand as an inherent sub-element necessary for any of the other causes to exist in the first place. Put another way, absent over-demand, there would just be "volume," but not necessarily "congested" volume.

conditions. Examples would be peak-hour slowdowns at junction points, intersections, and ramps. One can "plan" for these events because one knows by routine that such events will occur time and again in the same manner and place.

Traffic microsimulation tools – Complex microsimulation tools that rely on input of traffic data, intersection "nodes," facility "links," and the associated parameters of each input, in order to output simulated conditions. By changing the inputs, engineers can test different sizes, characteristics, and out-year scenarios of traffic demand.

Upstream traffic – Traffic that has not yet arrived at the subject point on a highway.

Traffic Bottlenecks – Localized sections of highway where traffic experiences reduced speeds and delays due to recurring operational conditions or nonrecurring traffic-influencing events

	Recurring – "Predictable" in cause, location, time of day, and approximate duration. **Nonrecurring** – "Random" (in the colloquial sense) as to location and severity. Even if planned in some cases, like work zones or special events, these occurrences are irregular and are not predictably habitual or recurring in location.
	Recurring: Operational Causes – A "facility determinate" condition wherein a fixed condition (the design or function of the facility at that point) allows surging traffic confluence to periodically overwhelm the roadway's physical ability (i.e., capacity) to handle the traffic, resulting in predictable periods of delay. **Nonrecurring: Dynamic Occurrences** – An "event determinate" occurrence, wherein a dynamic situation either reduces available capacity (e.g., loss of lanes due to incident or work zone) or increases demand (e.g., special event).
	Recurring: Ramps, lane drops, weaves, merges, grades, underpasses, tunnels, narrow lanes, lack of shoulders, bridge lane reduction, curves, poorly operating traffic signals. **Nonrecurring:** Work zones, crashes, incidents, special events, weather.
	"Active" bottlenecks – When traffic "released" past the bottleneck is not affected by a downstream restriction (i.e., queue spillback) from another bottleneck. "Hidden" bottlenecks – When traffic demand is metered by another upstream bottleneck(s); i.e., either a lesser or nonexistent bottleneck that would increase or appear, respectively, if only unfettered traffic could reach it.
	Motorists typically refer to bottlenecks in terms of added time delay when compared to the same nondelayed trip, but engineers and agencies also measure performance data: average speed (travel time), lane densities, queue lengths, queue discharge rates, vehicle miles of travel (VMT), and vehicle hours of travel (VHT).
	Data is collected using manual techniques (e.g., floating cars, aerial photography, or manual counts from video recordings) or from dynamic surveillance (e.g., detectors, radar, video, etc.) collected in real time. Modeling, especially microsimulation, can be used to study the impacts of bottleneck remediation on upstream and downstream conditions.

(Continued)

<table>
<tr><td></td><td colspan="3">Recurring: Type I – Demand surge, no capacity reduction (typically at freeway on-ramp merges). Type II – Capacity reduction, no demand surge (typically changes in freeway geometry; lane drop, grade, curve). Type III – Combined demand surge and capacity reduction (typically in weaving sections).
Nonrecurring: Usually classified by the type of event (e.g., incident, work zone) and severity of impact (e.g., duration of the number of lanes lost, closed, or impassable).</td></tr>
<tr><td></td><td colspan="3">Recurring: Bottleneck is due to over-demand of volume (i.e., peak-hour conditions). The bottleneck clears from the rear of the queue as volume declines.
Nonrecurring: Bottleneck is due to loss of capacity due to an incident, or short-term over-demand due to a spot event. The bottleneck clears from the front or rear of the queue, depending on whether the cause is incident-related (former) or volume-related (latter), respectively.</td></tr>
<tr><td rowspan="13">Practical Mitigations:</td><td>Recurring: Corridor Congestion</td><td>Recurring: Localized Bottlenecks</td><td>Nonrecurring</td></tr>
<tr><td>Dynamic pricing</td><td>Use shoulder lane</td><td rowspan="12">Improve incident response capabilities; reduce incident impact; reduce on scene time for clearing incidents; reduce facility “downtime” during the event. In work zones, maintain maximum number of open lanes during peak times; shorten durations using innovative methods and contracting practices; minimize number of times a section is an active work zone by combining improvements (e.g., paving and safety) and using highly durable materials; employ least intrusive detour(s).
Pre-plan for and coordinate special events to adequately and efficiently handle event traffic, including not only the main event but the subordinate deliver-ies, VIP access, emergency response, and pre- and post-event activities.
Have predetermined detour plans for particular sections of highway in the event of weather- or incident-related events, including available tools (i.e., arrows, sign stands, VMS boards, public information conduits, etc.).</td></tr>
<tr><td>Transit alternatives</td><td>Restripe weave area</td></tr>
<tr><td>Ridesharing, telecommuting</td><td>Improve merge area</td></tr>
<tr><td>High-occupancy lanes</td><td>Widen, extend, remove, or consolidate ramps</td></tr>
<tr><td>Successive ramp metering</td><td>Individual metered or signalized ramp</td></tr>
<tr><td>New construction</td><td>Improve signalization or intersection design</td></tr>
<tr><td>Install frontage roads</td><td>Install frontage road</td></tr>
<tr><td>Traffic demand management (TDM) techniques</td><td>Effect “speed harmonization” as in Europe</td></tr>
<tr><td>Build park-and-ride lots</td><td>Encourage “zippering”</td></tr>
<tr><td>“Downtown“ or cordon\ congestion pricing</td><td>Use access management techniques</td></tr>
<tr><td>Provide traveler information</td><td>Provide traveler information</td></tr>
<tr><td>Proactive signal timing plans (including adaptive control)</td><td></td></tr>
</table>

In: Traffic Bottlenecks
Editors: A. L. Sloan and C. F. Dorsey ISBN: 978-1-62081-180-1

Chapter 2

AN AGENCY GUIDE ON OVERCOMING UNIQUE CHALLENGES TO LOCALIZED CONGESTION REDUCTION PROJECTS*

United States Department of Transportation
Federal Highway Administration

EXECUTIVE SUMMARY

The Federal Highway Administration's (FHWA) Localized Bottleneck Reduction Initiative (LBR) program focuses attention on mitigating the operational causes of recurring congestion "hot spots" (i.e., traffic bottlenecks) at ramps, merges, lane drops, intersections, weaves, etc. One of the efforts of the LBR program is to encourage agencies to adopt a defined, "named" annualized spot-congestion program in the same manner that they might have an annualized spot-safety program for high crash locations. In the course of conducting state visits to "spread the gospel" of the tremendous benefits of clearing up one or more congested bottlenecks, the LBR staff has occasionally heard "push back" from some agencies that cite institutional or other barriers to enacting either individual projects or agency-wide programs. Examples of

---

* This is an edited, reformatted and augmented version of United States Department of Transportation, Federal Highway Administration Publication, FHWA's Bottleneck Reduction Initiative Program Office of Operations.

some of these barriers would be "we can't enact (these types of solutions) because (we feel) they violate firm safety design tenets or regulations;" or, "we can't undertake a spot-solution on a freeway absent having a vetted, adopted, twenty-year plan (or similar) already in place;" or, "how would such projects affect (our) Metropolitan Planning Organization's (MPOs) air quality, nonattainment status?" Conversely, the LBR staff has conducted state visits wherein these questions never came up; either signifying no such concerns, or success in overcoming them.

The main questions that this guidance helps an agency frame are:

1) What are the most common barriers and challenges with addressing localized congestion problems?
2) What are some case study examples that highlight how barriers and challenges were overcome?
3) What are some of the key factors in successful implementation of localized bottleneck projects?

This document was developed to provide guidance to state and local transportation personnel on how to overcome barriers and challenges to implementation of localized congestion relief projects. It presents and describes examples of institutional, design, funding and safety challenges that agencies face when trying to develop unique solutions to localized congestion problems. The document also presents ten detailed case studies of projects and programs that illustrate how to overcome common barriers and challenges. The case studies were chosen to highlight agencies that have implemented effective projects in a unique and praiseworthy fashion. The final section provides some high-level guidance and practical ideas on how to implement successful solutions to localized congestion problems based on experience and information gathered during this project.

1.0. INTRODUCTION

1.1. Purpose of the Guidance Document

This guidance document provides guidelines that can be used by state departments of transportation (DOTs) and local transportation agencies. The guidance document was developed based on best practices used by state and local agencies during the planning, design and implementation of localized congestion (a.k.a. bottleneck) relief programs and projects.

1.2. How to USE This Document

Target Audience

This document is designed for state, regional, and local transportation agencies and private consultants that are focused on mitigating operational causes of bottlenecks. These bottleneck locations include a wide variety of causes from poorly functioning merges/diverges to poor ramp spacing throughout an entire freeway corridor.

This document targets planners as well as traffic, safety, and design engineers, because bottlenecks need to be addressed in all phases of the project development process. Operations and maintenance staff will also find this document useful because it highlights innovative thinking and action by agencies on implementing projects to relieve localized congestion caused by bottlenecks. The document includes a series of case studies and some high-level guidance on developing and implementing successful projects.

Target Audience:

- Transportation
    - State
    - Regional
    - Local
- Consultants
- Planners
- Designers
- Operations and maintenance

Document Structure and Content

This guidance document includes the following sections:

- **Section 1.0 – Introduction.**
  This section contains background information on traffic bottlenecks and describes how the FHWA is addressing bottlenecks through their LBR Program.

- **Section 2.0 – Common Barriers and Challenges to Localized Congestion Projects.**
  This section provides information on common barriers and challenges associated with implementing localized congestion relief projects.
  The barriers and challenges are divided into four primary categories, including: institutional, design, funding and safety.
- **Section 3.0 – Case Studies of Successful Programs and Projects.**
  This section summarizes case study examples of successful bottleneck programs and projects across the United States, focusing on ten that provide valuable lessons on overcoming common barriers and challenges.
- **Section 4.0 – High-Level Guidance on Implementing Successful Projects.**
  This section includes some high-level guidance on developing and implementing successful bottleneck and localized congestion relief projects.

1.3. Background

Bottlenecks: A Definition

The FHWA estimates that 40 percent of all congestion nationwide can be attributed to recurring congestion (see Figure 1); some of it "mega" where entire regions or large facilities (e.g., interchanges or corridors) are overwhelmed by seemingly unceasing traffic demand and some of it "subordinate" – locations on the highway system where periodic volume surges temporarily overwhelm the physical capacity of the roadway.

Of this 40 percent, there has never been research to determine how much is attributable to subordinate locations.

During off-peak hours, the subordinate locations operate sufficiently and safely for the conditions.

These recurring "localized" bottlenecks are those encountered in our everyday commutes, and are characterized as being relatively predictable in cause, location, time of day, and approximate duration.

Nonrecurring congestion, on the other hand, is caused by random events such as crashes, inclement weather, and even "planned" events such as work zones and special events.

Traffic Bottlenecks:

Localized sections of highway where traffic experiences reduced speeds and delays due to recurring operational conditions or nonrecurring traffic-influencing events.

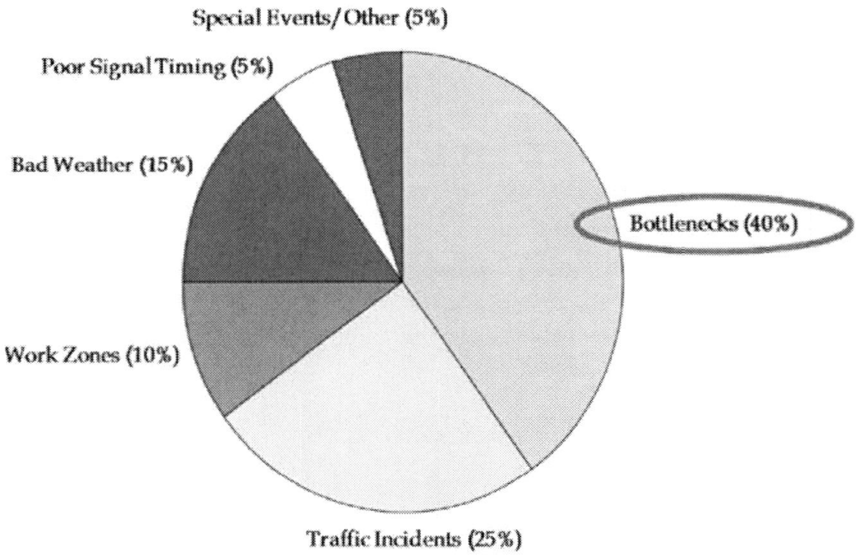

Source: FHWA-http://ops.fhwa.dot.gov/congestion_report/congestion_report_05.pdf.

Figure 1. Pie Chart. Sources of Traffic Congestion.

Localized Bottlenecks

This guidance document focuses on "localized" recurring bottlenecks (i.e., point-specific or short corridors of congestion). **Mega-bottlenecks or those occurring due to systemic congestion are not meant to be covered by this guidance.** It is understood that transportation agencies have different thresholds (financial and otherwise) of what it means to be a localized versus mega bottleneck project.

For a bottleneck to be "localized," the factors causing the bottleneck *ideally* should not exert influence upon, or be influenced by, any other part of the transportation system. As a practical measure, the LBR program recommends considering the closest upstream and downstream decision points as either impacting "to" or impacting "from" the subject location, respectively. Anything much beyond that reach might be considered more than "localized." One exception might be collector-distributor lanes that would almost certainly run through two or more on- or off-ramps. Such a "system" can be considered as a larger, localized condition. Otherwise, recurring, localized bottlenecks generally occur at the areas described in **Table 1**.

Localized Bottlenecks:

Usually exist in one direction (e.g., underserved movement)

- Predictable:

✓ Cause
✓ Location
✓ Time of day
✓ Duration

- Point-specific or short corridor
- Solutions to fix are:

✓ Small geometric changes
✓ Relatively low-cost
✓ Delivery is reasonably quick

(1 construction season or less)

Table 1. Common Locations for Localized Bottlenecks

Location	Symbol	Description
Lane drops		Bottlenecks can occur at lane drops, particularly mid-segment where one or more traffic lanes ends or at a low-volume exit ramp. They might occur at jurisdictional boundaries, just outside the metropolitan area, or at the project limits of the last mega project. Ideally, lane drops should be located at exit ramps where there is a sufficient volume of exiting traffic.
Weaving areas	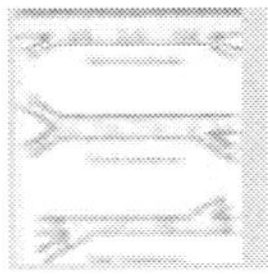	Bottlenecks can occur at weaving areas, where traffic must merge across one or more lanes to access entry or exit ramps or enter the freeway main lanes. Bottleneck conditions are exacerbated by complex or insufficient weaving design and distance.

Location	Symbol	Description
Freeway on-ramps		Bottlenecks can occur at freeway on-ramps, where traffic from local streets or frontage roads merges onto a freeway. Bottleneck conditions are worsened on freeway on-ramps without auxiliary lanes, short acceleration ramps, where there are multiple on-ramps in close proximity and when peak volumes are high or large platoons of vehicles enter at the same time.
Freeway exit ramps		Freeway exit ramps, which are diverging areas where traffic leaves a freeway, can cause localized congestion. Bottlenecks are exacerbated on freeway exit ramps that have a short ramp length, traffic signal deficiencies at the ramp terminal intersection, or other conditions (e.g., insufficient storage length) that may cause ramp queues to back up onto freeway main lanes. Bottlenecks could also occur when a freeway exit ramp shares an auxiliary lane with an upstream on-ramp, particularly when there are large volumes of entering and exiting traffic.
Freeway-to-freeway interchanges		Freeway-to-freeway interchanges, which are special cases on on-ramps where flow from one freeway is directed to another. These are typically the most severe form of physical bottlenecks because of the high traffic volumes involved.
Changes in highway alignment		Changes in highway alignment, which occur at sharp curves and hills and cause drivers to slow down either because of safety concerns or because their vehicles cannot maintain speed on upgrades. Another example of this type of bottleneck is in work zones where lanes may be shifted or narrowed during construction.
Tunnels/ underpasses	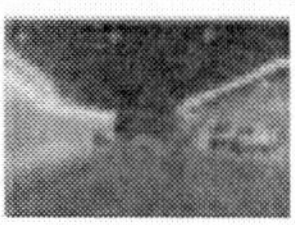	Bottlenecks can occur at low clearance structures, such as tunnels and underpasses. Drivers slow to use extra caution, or to use overload bypass routes. Even sufficiently tall clearances could cause bottlenecks if an optical illusion causes a structure to appear lower than it really is, causing drivers to slow down.
Narrow lanes/ lack of shoulders		Bottlenecks can be caused by either narrow lanes or narrow or a lack of roadway shoulders. This is particularly true in locations with high volumes of oversize vehicles and large trucks.

Table 1. (Continued)

Location	Symbol	Description
Traffic control devices		Bottlenecks can be caused by traffic control devices that are necessary to manage overall system operations. Traffic signals, freeway ramp meters, and tollbooths can all contribute to disruptions in traffic flow.

Table 2. describes five of the most common causes or reasons why localized bottlenecks exist.

Table 2. Common Causes for Localized Bottlenecks

Causal Factor	Symbol	Description
Traffic patterns change		Bottlenecks exist in freeway and arterial street systems because **traffic patterns change**, due to new development, changes in roadway characteristics (especially new roadways), employment, or other factors.
Traffic forecasting is inexact		Bottlenecks can exist because of the difficultly with forecasting growth in traffic volumes. **Traffic forecasting is inexact**, particularly when trying to look at peak hour and peak period travel characteristics.
Disconnect between disciplines	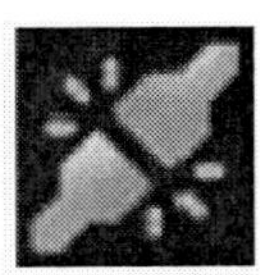	Bottlenecks can result from a **disconnect between disciplines** within the project development process, where communication among planners, design and operations personnel does not adequately address localized congestion concerns.
Lack of knowledge	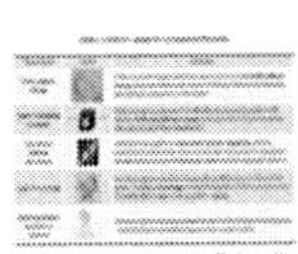	Sometimes agencies simply do not know where bottlenecks are located on their system. This **lack of knowledge** of localized congestion is often the result of a lack of available data to support good congestion mapping.
Misinterpretation: localized vs. systemic		Sometimes agencies **misinterpret that the situation is not localized but systemic** – characterizing the entire interchange or corridor as over capacity.

A detailed discussion on bottleneck characteristics is provided in FHWA Publication FHWAHOP-09-037, *Recurring Traffic Bottlenecks: A Primer – Focus on Low-Cost Operational Improvements*, available on FHWA's web site at http://www.ops.fhwa.dot.gov/publications/fhwahop09037/fhwahop09037.pdf.

Bottlenecks: A History

Timeline of National Bottleneck Activities

Over the past decade, transportation professionals have come to realize that highway bottlenecks demand special attention. Several national studies have highlighted bottlenecks as a major congestion problem in urban areas. These studies have raised the level of awareness about bottlenecks as a problem, warranting that they be treated as a significant part of the congestion problem. One of the LBR tenets is "a bottleneck is congestion, but congestion is not always just a bottleneck."

Figure 2. Graphic. The Endless Pursuit of Congestion Relief.

This means that a bottleneck (or chokepoint) is merely a subset of the larger congestion pie. However, that "subset" is now realized to be a uniquely impacting (and increasingly growing) genre of congestion; namely, that it is *subordinate* locations along a highway that need to be fixed, and not necessarily the knee-jerk expectation to rebuild the entire facility. Granted, in some cases, an aging or clearly capacity-deficient facility may need to be replaced. But in this age of budget constraints and economizing, one or two corrections to inefficient subordinate locations on a facility may be all that is needed to improve the condition. **Figure 2** takes a satirical approach to the argument that we can build our way out of congestion by continued major expansion of freeway facilities.

1.4. Localized Bottleneck Reduction Program

FHWA's Localized Bottleneck Reduction Program promotes operational and low-cost bottleneck mitigation strategies to improve mobility. Managed by the Office of Operations, the program serves to bring attention to the root causes, impacts, and potential solutions to recurring traffic chokepoints; ones that are wholly the result of operational influences. This is "good and bad" news in the sense that design influences can always be corrected, but some corrections may be cost-prohibitive in terms of direct construction costs or indirect right-of-way impacts. Regardless, many locations have the potential to be corrected for relatively low-cost and with relatively low physical impact. In any case, the goal of the program is to raise awareness of bottlenecks at the state level and promote low-cost, quick-to-implement geometric and operational improvements. The LBR Program has several activities either completed or underway, including:

- This guidance document, which provides guiding principles and concepts common to knowing and overcoming the barriers and challenges to implementing low-cost operational improvement programs and projects.
- A companion guidance document, *An Agency Guide on How to Establish Localized Congestion Mitigation Programs*, developed to provide guidance to state and local personnel who wish to develop a formal program for mitigating congestion using localized and low-cost treatments. It presents templates for developing a localized congestion mitigation program, including documenting alternative templates in use by state DOTs and Metropolitan Planning Organizations (MPO).
- *Traffic Analysis Toolbox Volume X: Localized Bottleneck Congestion Analysis – Focusing on What Analysis Tools are Available,*

Necessary, and Productive for Localized Congestion Remediation. This document provides guidance on tools required to analyze the specific genre of localized congestion problems.
- *Recurring Traffic Bottlenecks: A Primer – Focus on Low-Cost Operational Improvements.* This Primer is the "face" of the program. It provides an overview of the wide range of operational and low-cost strategies available to reduce congestion at bottlenecks.
- Localized Bottleneck Reduction Regional Workshops. Regional workshops for state and local agencies to learn and share information on localized bottleneck reduction strategies and how they can be incorporated into state and local planning processes.
- Many of the items listed above can be found at the FHWA bottleneck web site (http:iops.fhwa. dot.gov/bn/index.htm), which can be found at the FHWA Office of Operations web site.

Additional guidance documents are forthcoming that are aimed at agencies and personnel who have first responsibility to address bottleneck congestion locations.

2.0. Common Barriers and Challenges to Localized Congestion Projects

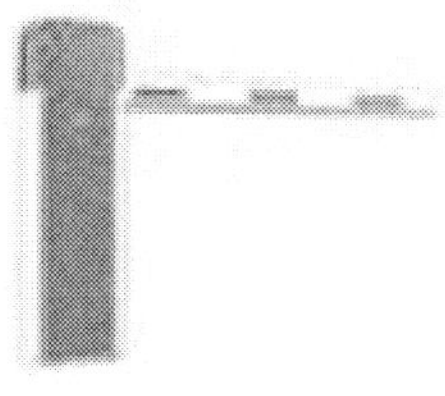

2.1. Background

There really are no set, widely utilized guidelines for roles and responsibilities of an LBR program or project development process. State DOTs, MPOs, or local transportation agencies could all lead an effective LBR effort or individual project. State DOTs and MPOs are traditionally the organizations who lead LBR efforts simply because they usually have larger missions, which include congestion management and mitigation, as well as access to a variety of funding mechanisms.

Many successful LBR programs actually depend on a high level of coordination between state DOTs and MPO. Many times, the state may identify bottlenecks and work closely with MPOs to integrate these projects into their Transportation Improvement Program (TIP) and other targeted funding sources such as Congestion Mitigation and Air Quality (CMAQ) and safety. However the split or leadership role is defined, any agency can lead an effective program.

This current document is aimed at determining and understanding the common barriers/ challenges to localized congestion mitigation programs and projects and why some agencies have had success overcoming them while others have not. The research team used a variety of methods to gather information on common barriers/challenges, including an agency survey, attendance at LBR workshops, and a state-of-the-practice review of published studies and internet pages dealing with the subject of localized congestion and bottleneck removal projects.

LBR Workshop Feedback

FHWA has sponsored a number of workshops to help partner agencies become engaged in starting or improving efforts aimed at reducing localized bottlenecks. LBR staff has occasionally heard "push back" from some agencies that cite institutional or other barriers to enacting either individual projects or agency-wide programs.

Examples of some of these barriers would be:

1) "we can't enact (these types of solutions) because (we feel) they violate firm safety design tenets or regulations"
2) "we can't undertake a spot-solution on a freeway absent having a vetted, adopted, twenty-year plan (or similar) already in place"
3) "how would such projects affect (our) MPOs air quality, nonattainment status?"

Conversely, the LBR staff has conducted state visits and workshops wherein these barriers never came up; either signifying no such concerns, or success in overcoming them. Those agencies that have been effective at dealing with bottlenecks have developed either special or ongoing programs specifically targeted at dealing with current bottleneck projects.

The options for structuring an LBR program vary widely, as described in *An Agency Guide on How to Establish Localized Congestion Mitigation Programs*. There is no cookie-cutter approach to implementing low-cost bottleneck projects.

2.2. Barrier/Challenge Categories

This section describes the four categories for common barriers and challenges to localized congestion and bottleneck reduction projects. The research team developed the categories based on synthesizing available information from the LBR workshops, agency surveys, and state-of-the-practice review. **Table 3** outlines the four categories for most of the common barriers and challenges to localized congestion and bottleneck reduction projects.

Table 3. Bottleneck Barrier and Challenge Categories

Category	Symbol	Description
Institutional		Barriers and challenges that are bureaucratic in nature – caused by the way transportation agencies have traditionally operated and functioned.
Design	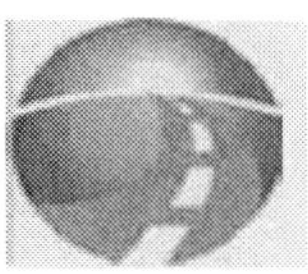	Barriers and challenges that are related to the way the transportation facilities are physically designed – particularly in relation to accepted standards and practices.
Funding		Barriers and challenges that are related to how transportation facilities are paid for and implemented.
Safety		Barriers and challenges that are related to the potential traffic safety impacts resulting from localized congestion relief treatments.

2.3. Institutional Barriers and Challenges

This section outlines the common barriers and challenges to localized congestion projects that are bureaucratic in nature – that is they are caused by the way agencies have traditionally operated. The research team identified ten common institutional barriers shown in **Table 4**.

Table 4. Common Institutional Barrier and Challenges

Barrier	Symbol	Description
Project champion	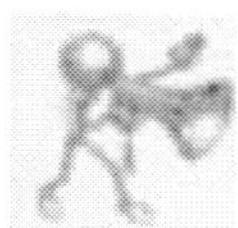	Localized congestion relief projects often need a project champion to be successful. Someone with a high-level of authority and ability to gain consensus make implementation easier.
Disposition towards mega projects		Many transportation agencies are organized to plan, design and construct large projects and do not have a well defined process for smaller, localized congestion reduction projects.
Project planning and programming requirements		Transportation agencies that receive federal funds use a defined planning process that is sometimes at odds with implementing localized congestion reduction projects (see air quality conformity later in this section).
Lack of training		There is a lack of available training for DOTs and MPOs on how to properly identify, analyze, and successfully implement localized congestion reduction projects.
Knowledge of problem locations		The barrier for some agencies is a lack of knowledge of localized congestion locations in their jurisdiction that might be mitigated with a low-cost/spot operational or geometric improvement.

Barrier	Symbol	Description
Deficiency with internal and/or external communication	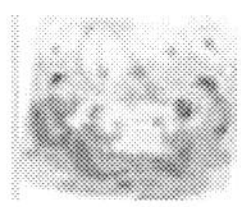	Communication – both internally within a transportation provider – and externally with partner agencies or key stakeholders can often be a barrier to project initiation and ultimately field implementation.
Culture of historical practices/resistance to change		Many DOTs struggle with organizational change and rely heavily on historical practices and approaches to project development.
Lack of incentives or recognition		Formal incentives and/or recognition for successful implementation of localized congestion reduction projects are not widespread.
Project is not in or consistent with approved regional and state transportation plans		Some agencies will not implement localized congestion reduction projects unless they have been through a formal process and been added to approved regional and state transportation funding plans.
Lack of confidence in proposed solution		Some engineers struggle with confidence in the ability of low-cost/localized improvements to be effective at reducing the congestion and not just moving it or making the situation worse.

Example of Institutional Barrier: Project Planning and Programming Requirements – Air Quality Conformity

An example barrier to localized congestion reduction projects in large urban areas relates to **air quality conformity**. Because they are short-term in nature, localized bottleneck improvements may emerge as formal projects that have not been previously identified in Statewide Transportation Improvement Programs (STIP) or TIP.

Thus, they may not be part of those projects that have been approved to deal with air quality issues in the region or state. Such occurrences must be dealt with on a case-by-case basis by agencies wishing to undertake bottleneck projects.

One point worth noting: if air quality conformity in a location precludes or discourages major capital expansion (e.g., additional lane-miles), the type of improvements in a localized bottleneck program clearly do not fall in this category.

Example of Institutional Barrier: Consistency with Long-Range Transportation Plans

Another example of a potential institutional barrier is that **bottlenecks may not be seen as consistent with long-range transportation plans**. Most bottleneck mitigation strategies such as roadway widening, left-turn lengthening, auxiliary lanes on freeways, or improvement of weave/merge areas may all be seen as distracting resources or blurring the need for larger design solutions, which will be made anyway in a larger longer-term project already in a 20- year plan.

Agencies must decide and weigh the benefits themselves whether the cost of doing smaller bottleneck solutions in the short term is against the cost of waiting for a more complete solution.

This decision can be difficult, especially for agencies without a good appreciation for the typical benefits and costs of smaller bottleneck solutions and how long those benefits might last.

2.4. Design-Based Barriers and Challenges

This section outlines the common barriers and challenges to localized congestion projects that are related to the way the transportation facilities are physically designed – particularly in relation to accepted standards and practices. The research team identified five common design-based barriers shown in **Table 5**.

Example Design-Based Barrier: Design Exception Process is Difficult

An example design-based barrier to localized congestion reduction relates to the FHWA **design exception** process. Because some bottleneck treatments use innovative solutions that maximize effectiveness with a minimum of new construction, they are occasionally at odds with highway design standards and might require a design exception (e.g., the addition of slip ramp to a collector/distributor road or the use of a shoulder as a through lane at selected locations may not strictly follow design standards).

Such deviations have the potential to degrade safety if not properly implemented (e.g., shoulder elimination may lead to more collisions with roadside features or may impede incident management activities).

As it is FHWA's intent to foster creative approaches for low-cost bottleneck projects, agencies should not see the design standard issue as insurmountable. Rather, they should fully assess the potential safety impacts of strategies and devise ways of addressing them, if necessary.

Table 5. Common Design-Based Barriers and Challenges

Barrier	Symbol	Description
Design exception process is difficult		Many bottleneck solutions require design exceptions for narrow lanes, shoulder width, etc. and the process of getting approval can sometimes be difficult.
Non-standard design is considered a deal-breaker		Many transportation agencies adhere strictly to design standards such as the AASHTO *Policy on Geometric Design of Highways and Streets* and do not move non-standard designs forward to implementation.
Problem is too big and nothing short of a total rebuild will fix it		A common thought process by DOTs, MPOs and elected officials is that most congestion problems require a large investment and multi-year construction to fix.
Spot treatment will move the problem and not fix it		Quite a few planners and engineers often believe that smaller, spot treatments will move the bottleneck and not fix it; therefore, the project is not pursued.
Standard practices contribute to bottleneck formation		Some agencies cite that existing design standards and practices (e.g., maintaining basic number of lanes through major freeway-to-freeway interchanges) actually contribute to the formation of bottlenecks.

For example, in the case of a shoulder-to-lane conversion, review of crash data, and the specific roadway location (perhaps through a Roadway Safety Audit), it may be determined that a barrier is required to keep vehicles off of the roadside. It may also require a change in incident management policy that would allow emergency vehicles to access incidents from the opposite direction. Finally, agencies should be in contact with the FHWA Division offices throughout the process as design review may be required, depending on circumstances.

2.5. Funding Barriers and Challenges

This section outlines the common barriers and challenges to localized congestion projects that are related to how transportation facilities are paid for and implemented. The research team identified four common funding barriers shown in **Table 6**.

Example Funding Barrier: Lack of Available Resources for Implementation

An example funding barrier to localized congestion reduction relates to the lack of available resources for implementation. Many transportation agencies are dealing with increased congestion and transportation needs while receiving less funding into the future. On a high level, most of the funding for roadway improvements comes from motor vehicle fuel taxes.

This revenue source is fixed (most state and federal fuel taxes have remained at the same level since the early 90s) and the buying power has been diminished by increasing construction costs and the increasing overall fuel efficiency of the vehicle fleet. All of this makes agencies careful about where money is spent. Some DOTs have taken steps to reduce expenditures by outsourcing operations and maintenance activities.

One example of a reduced in-house function that has affected their ability to implement localized congestion reduction projects is not having dedicated crews and equipment for roadway restriping. Simple restriping – such as adding marking to create option lanes at diverge points – is sometimes all that is needed in order to mitigate an existing bottleneck. Not having a dedicated striping crew or equipment has made it difficult to implement small, low-cost restriping projects because they are not cost efficient for their private contractors.

Table 6. Common Funding Barriers and Challenges

Barrier	Symbol	Description
There is no dedicated funding category or named program		Bottleneck projects typically do not have a dedicated funding category like what exists for safety. Unless there is a formal program identity, bottleneck remediation is usually relegated to a few project completed as part of an annualized safety program, or as a subordinate part of larger, other purposed projects.
Low-cost solution may blur or preclude the need for a bigger project	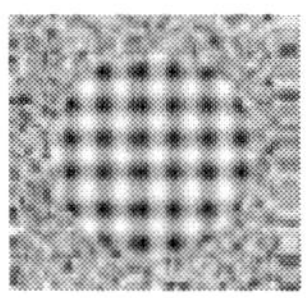	Some agencies do not implement localized solutions because they might distract resources from or blur the need for a bigger capital investment project.
Do not understand if alternative funding categories can be used		Transportation funding can be a complex process and some agencies do not understand what types of categories can be used to support implementation of localized bottleneck reduction-type projects.
Lack of available resources for implementation		Perhaps the most universal barrier is a lack of available resources for building, maintaining and operating transportation infrastructure. Some agencies cite lack of roadway striping crews as a challenge.

2.6. Safety-Related Barriers and Challenges

This section outlines the common barriers and challenges to localized congestion projects that are related to the safety performance of transportation facilities. The research team identified four common safety-related barriers shown in **Table 7**.

Example Safety-Related Barrier: Lack of Shoulders Takes Away Refuge Areas

An example safety-related barrier to localized congestion reduction relates to when an agency rejects moving a project forward because of concerns about the lack of roadway shoulders. The concerns about safety being compromised and incident management being more difficult, whether real or perceived, have to be adequately addressed before there is a comfort level to implement a project.

Transportation agencies can struggle with the paradox and balancing act of putting safety first with the implementing mobility solutions such as shoulder removal – even if it is for short sections like would be the case for localized congestion reduction projects.

Table 7. Common Safety-Related Barriers and Challenges

Barrier	Symbol	Description
Hesitancy to implement a solution that does not follow standards	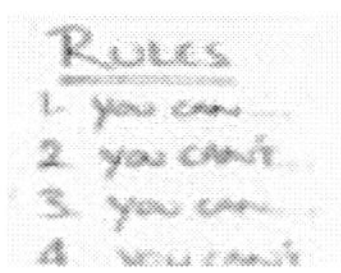	Agencies typically follow design standards and accepted practices very strictly in order to promote consistency and meet driver expectation. There is often a hesitancy to implement solutions that do not follow standards due to fear of unwanted outcomes.
Perception that safety is compromised with low-cost solutions		Some agencies have an organizational culture with a strong commitment to putting safety first and perceive that safety might be compromised by low-cost solutions, particularly when non-standard designs are involved.
Lack of shoulders takes away refuge areas		Utilization of shoulders as travel lanes, either permanently, during peak periods, or by special vehicles such as buses, can be an effective bottleneck improvement. Some agencies are reluctant to take away shoulders because the cross section is reduced and the refuge areas for disabled vehicles are eliminated.
Lanes that are not full width create safety issues for large trucks		Transportation agencies reduce lane widths in order to create an additional travel lane for bottleneck relief. Some agencies are adverse to this practice because of the potential safety implications – particularly when it is on freeways with high truck volumes.

3.0. CASE STUDIES OF SUCCESSFUL PROGRAMS AND PROJECTS

3.1. Overview of Case Study Process

Much of the emphasis in this document to this point has been on identifying and describing barriers and challenges to implementing localized congestion (bottleneck) programs and projects.

It is evident that while there is a wide range of barriers and challenges, many agencies have found ways to overcome them and implement programs and projects that are significantly successful and improving safety and mobility at relatively low cost.

This section focuses on highlighting a variety of case studies that illustrate key principles and lessons on how the barriers and challenges were overcome. The research team used an assortment of methods to gather information on the case study programs and projects, including an agency survey, attendance at LBR workshops, and a state-of-the-practice review of published studies and internet pages dealing with the subject of localized congestion and bottleneck removal projects.

Agency Survey

The research team conducted the research necessary to gather information for telling the story of the individual case-study examples that illustrate successful approaches and concepts for overcoming common barriers and challenges.

The research was accomplished using a variety of means, including:

1) Internet searches,
2) Phone interviews, and
3) Electronic surveys.

The research team designed a short survey instrument to gather details regarding approximately forty potential case study projects where agencies implemented localized congestion relief projects. The survey collected information on:

- **Project background:** roadway description, type of improvement, cost and implementation date.

- **Implementation barriers**: checklist of common barriers with opportunity to provide original ideas.
- **Strategies/methods for overcoming barriers**: how were individual barriers overcome and what was the most difficult.
- **Catalyst for project**: was there an event or circumstance that caused the agency to address the localized congestion with a low-cost reduction project?
- **Outcome assessment**: what was the result (performance measurement, community reaction, awards/recognitions, etc.) of the reduction project?

State-of-the-Practice Review

The research team also performed a state-of-the-practice review to gather information on potential case study projects and programs. Internet searches and review of published literature and sources generated during workshops and state visits were the key information sources.

3.2. Case Study Matrix

The information gathered online was synthesized into the *Overcoming Challenges Matrix* shown in **Table 8**, which lists a number of case study examples where the most common challenges were overcome.

The *Overcoming Challenges Matrix* provides one or more case study examples; however, detailed information was not available for all of these examples but a web link is provided in the outcome column in order to give readers a way to get additional information if it is desired.

This document also contains a number of detailed case studies that are provided in the following subsections of Section 3 and also in Appendix A.

3.3. Detailed Case Studies

Table 9 lists the case study sites selected to illustrate key principles and lessons on how the barriers and challenges were successfully overcome. Each of the selected projects and programs demonstrates a somewhat unique approach and has practical value for agencies that want to start or expand an effort aimed at mitigating localized congestion resulting from bottlenecks. These sites also had sufficient information available—including their background, barriers, improvement strategies, circumstances, and outcome assessment—that made a detailed case study possible.

The following subsections in Section 3 give a brief overview of the sites selected for detailed study. One-page summary fact sheets of the case studies are also provided in Appendix A.

Table 8. Overcoming Challenges Matrix[1]

	Challenge Description	Case Studies	Outcome
Institutional	Having a project champion	Dallas, TX Kansas City, KS	+: 20+ projects due to DOT/MPO champions +: Governor passes bill allowing buses on shoulders
	Disposition towards mega projects	Minneapolis, MN Manchester, NH	+: Similar benefit for $7 vs. $138 million projects +: Expedited work at Exit 5 as part of mega project
	Project planning and programming requirements	Danbury, CT Austin, TX	+: Restriping at Exit 7 improved flow significantly +: Multi-disciplinary group mitigating congestion
	Lack of training/understanding on how to develop a successful project	Dallas, TX LBR workshops	+: Freeway Bottleneck Workshop +: Federal outreach workshops building consensus
	Knowledge of problem locations that can be fixed with low-cost solutions	Phoenix, AZ Dallas, TX Little Rock, AR	+: Regional bottleneck study +: Aerial freeway congestion mapping +: Operation Bottleneck program by MPO
	A culture of historical practices	Saginaw, MI	+: Successful roundabout at I-75/M-81 interchange
	Deficiency with internal and external coordination (design/operations)	New York, NY	+: PFI functional groups
	Can't implement projects without being in approved regional/state plans	Rhode Island DOT	+: Creation of the Strategically Targeted Affordable Roadway Solutions (STARS) program
	No incentive or recognition for successful low-cost bottleneck reductions	Dallas, TX	+: Engineers performance evaluation includes bottlenecks
	Will the proposed solution work – lack of confidence	Florida DOT	+: Trial fix with cones made permanent with striping

Table 8. (Continued)

	Challenge Description	Case Studies	Outcome
Design	Design exception (DE) process is difficult	Pittsburgh, PA	+: New shoulder to avoid DE, Academy at I-279
Design	"Non-standard" design is considered a deal-breaker	Minnesota DOT	+: Creation of "flexible design" concept
Design	Problem is too big and nothing short of a rebuild will fix it	Plano, TX	+: Implement auxiliary lane on US 75 at SH 190
	Spot treatment will move problem downstream and not improve mobility	Renton, WA	+: SR 167 spot fix near Boeing reduces congestion
	Standard design practices contribute to bottleneck formation	Fort Worth, TX	+: I-20/SH 360 fix defies AASHTO basic lanes policy
Funding	There is no dedicated funding category for this type of project	Mississippi DOT Nebraska DOT	+: I-10 shoulder use after Katrina improves flow +: ITS funds for ramp gates to fix US 75 bottleneck
Funding	Low-cost solution may blur or preclude need for bigger project	Dallas, TX	+: I-635 early action doesn't stop $3B mega project
Funding	Don't understand if alternate funding categories can be used	Virginia DOT Ohio DOT	+: STARS program uses safety $ to target congestion +: Safety funds include congestion index
Funding	Lack of available resources (e.g., DOT striping crews) for implementation	Dallas, TX	+: District striping contract implements small fixes
Safety	Hesitancy to implement solution that does not follow standard design	Minnesota DOT	+: Mobility crisis from I-35 bridge collapse
Safety	Perception that safety is compromised with low-cost, non-standard fixes	Texas DOT	+: Average 35% crash reduction for 13 projects in TX
Safety	Lack of shoulders takes away necessary refuge areas	Arlington, TX	+: Crash reduction at SH 360/Division
	Lanes that are not full width create safety issues for large trucks	Dallas, TX	+: I-30 Canyon truck rollovers basically eliminated

[1]For more details on these case studies, visit the informational links found in Appendix B.

Table 9. Overview of Detailed Case Studies

Case Study	Key Lesson Learned
Arkansas: Operation Bottleneck Program	"Ask and they will tell" → Active public involvement
California: I-580/US-101 Connector Ramp Restriping	"It is amazing what some white paint can do" → Low-cost restriping can really improve mobility
Louisiana: US-90 near Louisiana Superdome in New Orleans	"Sometimes less is more: creating exit-only lanes" → Eliminating a through lane can improve traffic flow
Maryland: Gateway Signal Optimization in Baltimore City	"Removing stops means more go's" → Optimizing traffic signals can yield significant results
Michigan: I-75/M-81 Interchange Reconfiguration	"Michigan roundabout proves golden" → Innovative design saves $6 million and wins award
Minnesota: I-94 Lane Modification near Lowry Tunnel	"Providing options can smooth flow" → Providing drivers an option lane reduces congestion
Minnesota: TH-100 at St. Louis Park	"Smaller can sometimes equal bigger" → Small bottleneck fix has similar result to mega project
Missouri: I-44/Route 13 Diverging Diamond Interchange (DDI)	"The early bird gets the worm" → First DDI project tough to sell but worth the effort
Texas: Low-Cost Freeway Bottleneck Removal Projects	"Championing low-cost projects" → Many successful projects where champions were key

3.4. Arkansas Case Study

Metroplan MPO

Metroplan, the MPO for the Little Rock region, has implemented a program dubbed "Operation Bottleneck" aimed at identifying current congested locations that are amenable to relatively quick and inexpensive treatments.

Major congestion problems – arterial corridors and freeway sections/interchanges with major capacity deficiencies – are well known throughout the area.

Further, future (major) problems have been identified with the modeling done for the long-range transportation plan.

However, funding for the major improvements necessary at these locations must come from either:

- State DOT, Arkansas State Highway and Transportation Department (they would be managed as state projects; and
- Local governments saving up several years of state and Federal allocations for a single project.

Metroplan wanted a way to serve their constituents better than constructing a scarce few megaprojects. Further, the region is almost in nonattainment for the eight-hour ozone standard. The text from their press release on the program sums up their intent very well: "We're aware of the major congestion issues in our area and have identified those in our long-range plans, but we know there are dozens, maybe hundreds of neighborhood problems throughout the region that could be fixed with something as simple as a roundabout or coordinating traffic signals to improve flow," McKenzie says. "Localized problems like these can be harder to identify and are sometimes overlooked, even though they can be just as frustrating to drivers. Often they can be addressed much more quickly than larger projects. Those are the types of areas we are hoping to identify through Operation Bottleneck."

Operation Bottleneck Program

Operation Bottleneck is largely based on the establishment of a Regional Mobility Authority (RMA). In Arkansas, an RMA is a coordinating body with no taxing powers – member counties would have to raise the taxes necessary to fund projects; multiple counties would be involved. Most likely, the RMA will be based on a temporary increase in local sales taxes county-by-county; they feel it is important to sunset the tax so it is more palatable to the public and elected officials. Metroplan hopes to leverage state and Federal funds against their self-generated revenue to fund the projects. Also key to the strategy is a specific list of projects to be funded by the tax increase, and most of the Operation Bottleneck effort has gone into project identification, as discussed below.

Project identification is being driven almost exclusively by public input via local meetings and an Internet survey. Metroplan also hired a marketing firm to promote the program through local media. Both congestion and safety problem areas are being solicited, along with other modal deficiencies (transit, special transportation). A huge range of responses has been received, from megaprojects to minor problems on local roads. For congestion problems, signals and interchanges are dominating the responses. Safety problems identified by the public tend to be more general than site-specific. (This is understandable since congestion is experienced routinely but crashes are rare events for individuals.)

Metroplan staff will assemble the projects and will develop a list of projects to iterate with the public. Staff will also make revenue projections under different sales tax rates. No formal benefits assessment is planned – as with project identification Metroplan emphasized that public input is the driver for Operation Bottleneck, not technical processes (which they use for all other transportation planning activities).

The staff will compare public-identified projects with those in the TIP and LRTP as well as against congested sections identified in their Congestion Management System in developing a prioritized list. Metroplan staff offered two types of improvements that are likely to dominate the project list:

- Low-cost arterial improvements – improved timing, intersection approach geometric improvements, and access management; and
- Roundabouts at uncontrolled, stop sign-controlled, or low volume signal locations.

Initial results were presented to public officials in October 2008. Metroplan would like to make this an ongoing process, especially since the public support for the program has been very high. How to structure the funding for an ongoing program will be tricky, however. Metroplan staff highlighted a $1.5 million dollar improvement project implemented in 2010 that alleviated intersection congestion at Dave Ward Drive and Donaghey Road in Conway, Arkansas as an effective outcome of the Operation Bottleneck program.

Table 10 outlines the key barriers and challenges and how they were overcome.

Table 10. Synopsis of Operation Bottleneck – Dave Ward Drive/Donaghey Road Case Study

Barriers and Challenges	How They Were Overcome
Other – congestion at intersection located near college, major intersection improvements proposed required additional right-of-way	Project in part associated with proposed development
Other – federal funds were first proposed for project, funds were shifted and all local money was used on the project	City identified local funding to be used on the project, with the limited federal funding being placed on another project that included federal funds

For More Information

See the fact sheet in Appendix A for a synopsis of the Operation Bottleneck case study.

3.5. California Case Study

Caltrans

The California Department of Transportation (Caltrans) does not have a formal bottleneck planning process; rather, bottleneck issues are addressed at the district level as part of their Corridor System Management Plans (CSMP), which are developed for some of California's most congested transportation corridors. System monitoring and evaluation is seen as the foundation for the entire process because it cannot only identify congestion problems, but also be used to evaluate and prioritize competing investments. The CSMP includes the identification of bottlenecks and potential short-term fixes as part of an overall and long-term strategy for making corridor improvements. This may take the form of an "LBR audit," which is a review of traditional large-scale corridor studies to identify opportunities for using LBR improvements as part of the package of improvements. The LBR audit concept is similar to that of Road Safety Audits. Caltrans does not have a direct funding for bottlenecks, although bottleneck projects are routinely programmed through the CSMP process.

I-580/US-101 Connector Ramp Restriping

The second detailed case study is a project in California that involved restriping a ramp to improve traffic flow. **Table 11** outlines the key barriers and challenges and how they were overcome.

Table 11. Synopsis of I-580/US-101 Connector Ramp Restriping Case Study

Barriers and Challenges	How They Were Overcome
Project champion	Public became champion due to organized outcry
Lack of confidence in proposed solution	Confidence from knowledge that this was a temporary solution.
Low-cost spot solution may blur the need for larger project	Still went forward with $10 million project for permanent widening of the ramp to two lanes
Lack of shoulders takes away necessary refuge areas	Loss of shoulders only a temporary condition

For More Information

See the fact sheet in Appendix A for further details of the I-580/US-101 Connector Ramp Restriping case study.

3.6. Louisiana Case Study

Louisiana Department of Transportation and Development

The Louisiana Department of Transportation and Development (DOTD) does not current have a formal localized bottleneck reduction program for project implementation. Low-cost projects that address localized congestion are done on an ad hoc basis within the various district offices.

US-90 near Louisiana Superdome in New Orleans

The fourth case study is a project implemented by the Louisiana DOTD in the New Orleans area on US-90, also commonly referred to as the Ponchartrain Expressway (PE). This project is unique because it is a case where DOT officials decided to eliminate through lanes on the PE to create exit-only lanes designed to reduce last-second merging and ease the chronic bottleneck. **Table 12** outlines the key barriers and challenges and how they were overcome.

For More Information

See the fact sheet in Appendix A for a synopsis of the US-90 near Louisiana Superdome in New Orleans case study.

Table 12. Synopsis of the US-90 near the Louisiana Superdome in New Orleans Case Study

Barriers and Challenges	How They Were Overcome
Lack of confidence in the proposed solution	DOTD staff performed a before study to get traffic volumes and determine that the reduction from 3 to 2 through lanes was not going to create another bottleneck in order to ease fears of management staff
Spot treatment will move the problem and not fix it	DOTD staff agreed to perform an after study to assess the operational outcome and make sure the bottleneck did not just move downstream

3.7. Maryland Case Study

Maryland State Highway Administration

The Maryland State Highway Administration (SHA) has a dedicated program of about $5M per year for the identification and implementation of low-cost traffic congestion improvements at intersections. The program's genesis tracks to when SHA asked, "What can be done if and when a megaproject's 'no-build' alternative is chosen?" The program has been well received by the public and local governments. Projects typically include low-cost projects that can be implemented quickly, such as signal timing upgrades and adding turn lanes and through lanes at intersections. The Maryland SHA also has had considerable success with projects to improve freeway ramps and merge areas that have reduced congestion bottlenecks at a low cost.

Baltimore City Gateway Signal Optimization

The fifth case study is a project implemented in Baltimore City, Maryland involving optimization of traffic signals in nine regionally significant arterial corridors. The signal retiming was implemented for less than half a million dollars and produced an overall benefitto-cost ratio of 51:1 based on benefits accrued from reductions in vehicle delay, number of stops, fuel consumption, and particulate emissions. Table 13 outlines the key barriers and challenges and how they were overcome.

For More Information

See the fact sheet in Appendix A for a synopsis of the Baltimore City Gateway Signal Optimization case study.

Table 13. Synopsis of the Baltimore City Gateway Signal Optimization Case Study

Barriers and Challenges	How They Were Overcome
Lack of training	City staff hired a consultant with experience in corridor signal timing projects
Culture of historical practices	Prior to the project the City signal shop had a lot of control over signal timings; however, majority control was switched to engineering
No dedicated funding category	City obtained funding using the CMAQ program in additional to local funds
Lack of available resources	Hiring a consultant supplemented available City staff

3.8. Michigan Case Study

Michigan DOT

Michigan DOT currently is in the process of developing a structured Localized Bottleneck Reduction (LBR) Program. The effort began several years ago with structured changes at MDOT, during which MDOT officially reorganized their Maintenance and Traffic and Safety Divisions to create a Division of Operations. The next step was the formation of a new section titled

Systems Operations and Management (SOM). One of their early charges was to develop an approach to identify and eliminate bottlenecks throughout the State. Several years previous to this reorganization, MDOT developed and utilized a "Choke Point" Program, and their current efforts are patterned after that effort.

One of the first official action steps that the SOM Section pursued was to solicit potential bottleneck locations and problem descriptions from each of their seven region offices. More than 200 locations were identified, with about one-third being freeway interchanges. Based on further review by the SOM Section, the total number of potential locations was reduced to approximately 125 locations, which they believed: 1) met their definition of a "bottleneck" location; and 2) had a potential cost-effective solution that could address the problem. One of the primary goals of this highly focused initial effort is to develop a documented and sustainable approach that can demonstrate excellent benefit-to-cost ratios, as well as justification for allocation and expenditure of funds on the statewide LBR Program. The underlying goal was to obtain leadership support and a dedicated funding template specifically for bottleneck reduction projects, which has now been achieved.

Many challenges exist as the Program and structure move forward. One primary challenge is the need to complete a detailed analysis necessary for a large number of potentially competing projects, as well as a freeway analysis of these projects. MDOT staff resources are limited and MDOT is reviewing the potential use of consultants and/or universities for project analysis. Another issue is how to justify and evaluate the impacts of the suggested changes as well as the existing problem. The intent is to create a level playing field for application of LBR funding by each of the seven regions. The third major challenge is the availability of funding. Michigan is going through an extremely dynamic period with the overhaul of the automobile industry, and their funding has been reduced. These issues are all being discussed and debated as MDOT moves forward to establish and document a formal, fully funded LBR Program.

I-75/M-81 Interchange Reconfiguration

The sixth case study is a project implemented in Saginaw, Michigan involving a reconfiguration of the I-75/M-81 interchange from a diamond to a modern roundabout. Given limited resources, the Michigan DOT chose to use an innovative design approach with roundabouts replacing the tight diamond. **Table 14** outlines the key barriers and challenges and how they were overcome.

Table 14. Synopsis of I-75/M-81 Interchange Reconfiguration Case Study

Barriers and Challenges	How They Were Overcome
Culture of historical practices	Local DOT staff had a focused discussion with management on the design concept – and the approximately $6 million dollar cost savings
Lack of confidence in proposed solution	Used a microscopic simulation model to analyze the roundabout solution and show a level-of-service (LOS) improvement from D to A
	Performed presentations to stakeholders and used an aggressive public
Other: getting stakeholder buy-in	involvement and education campaign including the use of special brochures, videos and newspaper editorials.

For More Information

See the fact sheet in Appendix A for a synopsis of the I-75/M-81 Interchange Reconfiguration case study.

3.9. Minnesota Case Studies

Minnesota DOT

Minnesota DOT was originally driven to explore low-cost congestion relief projects because of budgetary restrictions, but soon realized that these projects could be implemented very quickly and, as a bonus, were highly visible and popular with the public. In less than one year, the Minnesota DOT developed a highly accelerated process for bottleneck identification and prioritization, which led to many effective projects in the following two years. The Minnesota DOT also found that because of lower costs, it could identify multiple locations throughout the region and "spread around" bottleneck projects in an equitable way. This process consisted of a study that used a five-step process to narrow potential projects into a recommendation list to the state legislature. Evaluation of completed projects produced high benefit/cost ratios, usually greater than 8:1.

Note: Circa 2009, this one-time activity was replaced by an ongoing CMS process known as the Congestion Management Planning Process, which has been formally adopted as part of the 3C planning process.

Minnesota's Process to Identify and Prioritize Bottleneck Improvements

Step 1: Project Identification

Potential congestion management projects were identified from existing sources:

- Low-cost capacity improvements (e.g., auxiliary lanes);
- Restriping lane configuration; and
- Traffic control device improvements (e.g., ramp meters and signal timing).

Step 2: Quantitative Screening

- Project cost < $15 million
- Not in three-year TIP
- Annual hours of delay > 25,000
- Minimum of two hours of congestion

Step 3: Qualitative Screening

- Design readiness
- Cost range
- Congestion benefit
- Construction traffic management
- Future demand changes
- No adverse downstream effects

Step 4: Expert Workshop

Projects were prioritized by an expert group during a half-day workshop.

Step 5: Project Planning

The following were prepared for each project:

- Geometric sketches;
- Project scope;
- Congestion impacts;
- Safety impacts; and
- Benefit-to-cost ratio.

I-94 Lane Modification near Lowry Tunnel

The seventh case study is a project implemented in Minneapolis, Minnesota involving a lane modification near the Lowry Tunnel. The Minnesota DOT implemented a relatively low-cost ($300,000) project to provide an option lane on westbound I-94 approaching the I-394 interchange in proximity to the Lowry tunnel.

This is a relatively recent project (September 2010) that is still being assessed by DOT staff. **Table 15** outlines the key barriers and challenges and how they were overcome.

Table 15. Synopsis of I-94 Lane Modification near Lowry Tunnel Case Study

Barriers and Challenges	How They Were Overcome
Culture of historical practices	Education on benefits in short-term goals.
Lack of confidence in solution	Review of traffic analysis and other data supports the solution.
Spot treatment won't fix the problem	Commitment to monitor for one year and make necessary changes with future programmed project one year out.
Hesitancy to implement non-standard	Review of standards and how project does not violate driver expectation
solution	and, where design exception needed, determine likely outcomes.

TH-100 at St. Louis Park Project

The eighth case study is a second project implemented in Minneapolis, Minnesota involving improvements to a section of Trunk Highway 100 (TH-100) between 36th Street and I-394. The Minnesota DOT implemented a $7.1 million dollar project to add a third travel lane in one of the metro areas worst bottlenecks by converting roadway shoulders.

This project was completed in 2006 and then won an award for Public Project of the Year in Minnesota because of the significant mobility improvement. **Table 16** outlines the key barriers and challenges and how they were overcome.

For More Information

See the fact sheets in Appendix A for a synopsis of the I-94 Lane Modification near Lowry Tunnel and TH-100 at St. Louis Park Project case studies.

Table 16. Synopsis of TH-100 at St. Louis Park Project Case Study

Barriers and Challenges	How They Were Overcome
Culture of historical practices	Education on benefits in short-term goals.
Lack of confidence in solution	Review of traffic analysis and other data supports the solution.
Spot treatment won't fix the problem	Commitment to monitor for one year and make necessary changes with future projects.
Hesitancy to implement non-standard solution	Review of standards and how project does not violate driver expectation and, where design exception needed, determine likely outcomes.

3.10. Missouri Case Study

Missouri DOT

1. The Missouri DOT does not currently have a formal localized bottleneck reduction program.

Low-cost projects that address localized congestion are done on an ad hoc basis within the various district offices – most notably in the Springfield and St. Louis metropolitan areas.

The Missouri DOT utilized a "practical design" concept in the fall of 2005 that challenged internal staff, the FHWA and consulting community to help cut the budget of the 5-year STIP by 10 percent.

Engineers were told that they could put their design manuals on the shelf for a year and be guided by the following three rules:

1) **Safety** – every project must get safer. There is no room for compromise where safety is concerned.
2) **Communication** – there is collaboration in developing practical solutions.
3) **Quality** – the practical solution must function properly and cannot leave a legacy of maintenance challenges.

The District challenge resulted in an initial savings of over $400 million across the 5-year STIP. District representatives were assembled to discuss their experiences with practical design – both good and bad.

About 400 ideas and comments were discussed and documented and then boiled down to 25 broad policies in 5 general areas. These five areas accounted for 80 percent of the Missouri DOTs program delivery expenditures:

- Paving and based – 35 percent;
- Bridges – 17 percent;
- Grading – 11 percent;
- Right-of-way – 10 percent; and
- Traffic control – 7 percent.

The switch to the practical design concept in Missouri has produced a significant organizational change and also positive results in the safety, communication and quality as shown in **Table 17**.

Table 17. Outcome of Missouri DOT Organizational Change to Practical Design Concept

Category	Outcomes
Safety	• Largest drop in traffic-related fatalities of any state in the nation in 2006
	• Fatal crashes dropped below 1,000 in 2007 and still further in 2008.
	• 11% decrease in run-off-road crashes since 2004
Communications	• 90% of newspaper editorials in 2008 were positive
	• Customer satisfaction with Missouri DOT rose to 78% in 2008
	• 95% of customers believe projects are the right transportation solution
Quality	• Since 2002, Missouri DOT has delivered a $7 billion dollar program 0.4% under budget

I-44/Route 13 Diverging Diamond Interchange

The ninth case study is a first of its kind implemented in Springfield, Missouri. The Missouri DOT implemented a $3.2 million dollar project to convert an existing congested interchange (I-44/Route 13) into the first Diverging Diamond Interchange (DDI) in the United States.

The first DDI was somewhat of a battle to get in place but its success is leading to more implementations in Missouri.

Table 18 outlines the key barriers and challenges and how they were overcome.

For More Information

See the fact sheet in Appendix A for a synopsis of the I-44/Route 13 Diverging Diamond Interchange case study.

Table 18. Synopsis of I-44/Route 13 Diverging Diamond Interchange Case Study

Barriers and Challenges	How They Were Overcome
Project champion	The project engineer became champion and garnered by internal (upper DOT management) and external (business) support.
Project planning and programming requirements	The FHWA required that a modified Interstate Access Justification (IAJ) be done 2 weeks before the project was let; however, project staff was able to finish the new analysis prior to the deadline.
Lack of confidence in the proposed solution	Project staff utilized an enhanced level of traffic analysis and public outreach (particularly with Wal-Mart) including a simulation model and project video designed to showcase how a DDI operates.
Hesitancy to implement solution that does not follow standards	The project engineer had to convince DOT Central Design staff that the pedestrian access would work well through the middle of the DDI instead of on the outside like traditional diamonds.

3.11. Texas Case Study

Texas DOT

The Texas DOT does not currently have a formal localized bottleneck reduction program. Low-cost projects that address localized congestion are done on an ad hoc basis within the various district offices – most notably in Austin (see **Table 19**) and Dallas/Fort Worth (see **Table 20**).

Texas Low-Cost Freeway Bottleneck Removal Projects

The tenth case study is of a low-cost freeway bottleneck removal program primarily in the DFW metropolitan area. The Texas DOT implemented a number of low-cost projects to address localized congestion. Most projects utilized improvements such as restriping, shoulder conversion, and installation of auxiliary lanes to improve mobility in short sections of freeway.

Most improvements were implemented with local DOT discretionary funds, and projects champions at both the DOT and MPO were a key to success. **Table 21** outlines the key barriers and challenges and how they were overcome.

Table 19. Summary of Seven Bottleneck Removal Projects in Austin, Texas

Project Location	Implemented Improvements
I-35 NB at Parmer Lane	Supplemental lane was added from the Parmer Lane entrance to the Dessau exit and then extended to Wells Branch Parkway.
I-35 SB at Wells Branch Parkway	Closed Dessau Road entrance to southbound (SB) I-35 and added an auxiliary lane from the Wells Branch entrance to the Parmer Lane exit.
I-35 NB at US 183	Added an auxiliary lane from US 183 to the Braker Lane exit ramp.
I-35 SB at US 183	Began a 4th lane for SB I-35 at the Rundberg entrance (instead of the US 183 direct connector) and extended the auxiliary lane from the US 183 frontage entrance to the US 290 exit upstream so that it begins at the direct connector. The US 183 entrance from the SB I-35 frontage road was closed in order to facilitate flow along the auxiliary lane.
I-35 SB at Riverside Drive	Added an auxiliary lane from the Riverside Drive entrance to the Oltorf Road exit.
Loop 1 SB at Far West Boulevard	Added an auxiliary lane from the Far West Boulevard entrance to the Ranch-to-Market 2222/ Northland exit ramp.
Loop 1 at Loop 360	Realigned the SB Loop 1 main lanes so that the lane drop occurs at Loop 360 East rather than at the Loop 360 West exit ramp. Upstream of this lane drop are the high-volume Bee Caves Road entrance and the low-volume Barton Skyway entrance ramp to SB Loop 1.

For More Information

See the fact sheet in Appendix A for a synopsis of the Low-Cost Freeway Bottleneck Removal Projects case study.

Table 20. Evaluation of Thirteen Texas Bottleneck Projects – Mostly in Dallas-Fort Worth

District	Freeway(s) and Limits	Improvement Type	Annual Benefit	Cost	B-C Ratio*
FTW	NB SH360 @Division (SH180)	Shoulder conversion (outside) + auxiliary lane addition	$200,000	$150,000	10:1
ELP	EB I-10 @ US54	Restriping + ramp modification + auxiliary lane addition	$1.3M	$530,000	20:1
DAL	EB I-30 I-35E to I-45	Ramp reversal (exit converted to entrance) + auxiliary lane addition	$700,000	$660,000	9:1
DAL	NB I-35E I-30 to Dallas North Tollway	Shoulder conversion (inside) + auxiliary lane additions	$600,000	$130,000	37:1
DAL	EB SH190 to SB US75	Restriping + ramp modification	$500,000	$11,000	374:1

District	Freeway(s) and Limits	Improvement Type	Annual Benefit	Cost	B-C Ratio*
DAL	NB I-35E ramp to Dallas North Tollway	Re-striping + ramp modification	$300,000	$20,000	132:1
DAL	NB-SB I-35E Loop 12 to I-635	Shoulder conversion (inside) + removal of two inside merges	$11.0M	$1.9M	47:1
DAL	WB I-30 ramp to SB I-35E	Restriping + ramp modification	$200,000	$5,000	324:1
FTW	EB I-20 to NB SH360	Restriping + ramp modification + removal of thru lane inside interchange	$500,000	$10,000	400:1
FTW	SB SH360 to WB I-20	Restriping + ramp modification + removal of thru lane inside interchange	$30,000	$8,000	32:1
FTW	SB SH360 @ Division (SH180)	Ramp closure + auxiliary lane addition	$1.0M	$440,000	18:1
DAL	EB I-635 to NB US75	Restripe and widen left-side ramp from one to two lanes	$3.6M	$2.45M	24:1
DAL	SB US75 to WB I-635	Shoulder conversion (inside) on I-635 to allow ramp from US75 its own lane	$3.8M		

* B-C ratio based on ten-year project life with a 4 percent discount rate.

Table 21. Synopsis of Texas Low-Cost Freeway Bottleneck Removal Projects Case Study

Barriers and Challenges	How They Were Overcome
Project Champion	Support for low-cost bottleneck projects was at a high level, including the District Engineer and MPO Director.
Project planning and programming requirements	Most solutions were over a short distance where air quality and environmental requirements were not necessary.
Lack of training	The Texas Transportation Institute developed a Freeway Bottleneck workshop that taught the DOT about successful bottleneck removal.
Lack of incentives/recognition	DOT leadership formalized recognition for successful projects, rewarding Area Engineers during annual performance evaluations.
Design exception process is difficult	Over time, DOT gained the trust of FHWA staff by evaluating projects and showing the positive operational and safety benefits.

4.0. Guidelines for Successful Localized Bottleneck Reduction Projects

4.1. Overview of Guidance

This document, *An Agency Guide on Overcoming Unique Challenges to Localized Congestion Reduction Projects*, provides guidance developed from successful localized bottleneck reduction programs and projects. As has been demonstrated throughout this document, there is a wide variety of barriers and challenges to successful project implementation, and many agencies have taken unique approaches to overcome them. There is no one size fits all or cookie-cutter way of attacking localized congestion with low-cost bottleneck improvements. However, the following guidance is synthesized from practical experience from over 50 successful bottleneck removal projects and feedback from transportation professionals throughout the country.

4.2. Guidance for Successful Localized Bottleneck Reduction Projects

This section presents ten high-level guidelines to assist agencies in the development and execution of successful localized bottleneck reduction projects. Each of the ten guidelines is presented in a text box and then followed by a short explanation of how this guidance can be applied.

Guideline 1: Get a high-level project champion to promote the project and/or program

Having a project champion, ideally a high-level person with decision-making authority and respect from peers is desirable for getting low-cost localized congestion projects implemented. The tenth case study in Texas highlights the importance of this principle with champions from both the Texas DOT Dallas District Engineer (highest level DOT local authority) and the Director of Transportation for the North Central Texas Council of Governments, the Dallas/Fort Worth MPO.

Guideline 2: Find the exact problem and place the improvement to help this movement

It is critical to successful bottleneck removal to find the exact problem causing the congestion and place the right improvement *to help serve that underserved movement.* Resurrecting Transportation System Management (TSM) thinking is the key to use of this strategy of obtaining short-term congestion relief without unnecessarily impacting other movements or just moving the bottleneck to a nearby downstream location.

Guideline 3: Collect enough of the right data in order to perform the proper analysis

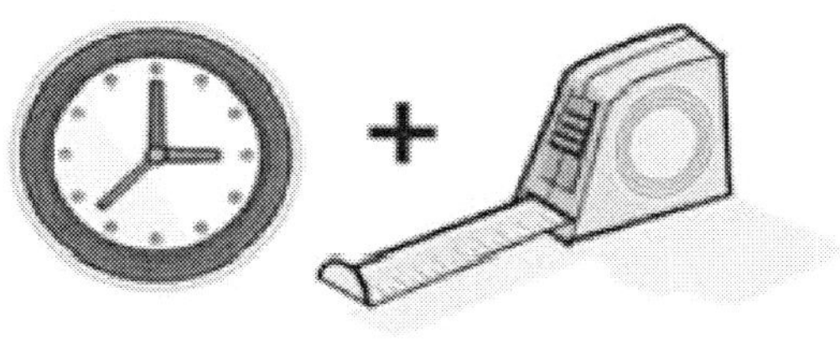

Another important component is to collect enough of the right data to capture the extent of queuing, both in time and space, so that a proper analysis of improvement alternatives can be performed. Having the right data before and after will be necessary for an appropriate performance evaluation to capture the real-world benefits after the project has been implemented. At times, the collection of speed data on parallel routes is important, since traffic diversion may occur after the improvement.

Guideline 4: Establish low-cost bottleneck removal in a formal project development process

It is important to realize that there is significant value in making low-cost bottleneck removal part of an established project development and/or planning process and involving multiple agencies in implementation. Examples of this type of approach and coordination were evident in the case studies presented in Section 3 and a companion guidance document, *An Agency Guide on How to Establish Localized Congestion Mitigation Programs*, which provides three templates that can be used to start new programs. Having a named program can also help gain visibility and support and lead to more frequent opportunities for successful project implementation.

Guideline 5: Use a team-based approach

Like many types of projects, implementation of localized congestion removal projects benefits from the use of a team-based, collaborative approach. Case studies of successful programs and projects have shown that many agencies emphasize interagency coordination (sometimes through formal processes such as traffic management teams, design charrettes, major investment studies, etc.) because of the positive synergy and opportunities for collective brainstorming. One project champion that was interviewed during this project indicated that "bottlenecks are a team sport".

Guideline 6: When possible test trial a bottleneck improvement prior to the permanent fix

Several case studies have demonstrated that it is a great advantage to be able to test a bottleneck improvement prior to the permanent fix. This allows an agency, and possibly even the public, to overcome fears about potential negative outcomes. The best case is being able to physically test the bottleneck improvement at the real-world site.

Examples of real-world trials include the Florida DOT using the cones for a temporary merge pattern prior to restriping (see Figure 3) and the Texas DOT using barrels to temporarily close ramps prior to their permanent closure. When a real-world trial is not feasible, the use of microsimulation models can serve the same purpose and also provide reassurance to both internal and external audiences about the likelihood of success of the proposed improvement.

Figure 3. Graphic. Florida DOT Tests New Merge Pattern Prior to Permanent Implementation.

Guideline 7: Utilize a consistent approach to project analysis and development

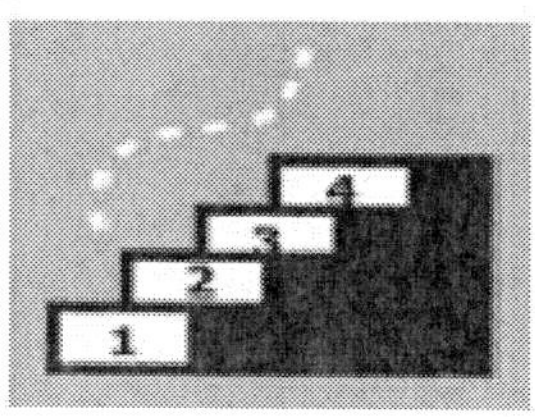

Another key component to successful project delivery is to utilize a consistent approach to project analysis and development.

The Texas Transportation Institute (TTI) has developed the four-step process, IDEA, based on analysis of dozens of freeway bottleneck projects.

The IDEA process consists of the following four steps:

1) *Identification* – screening of potential bottleneck locations.
2) *Detection* – finding the causes of the localized congestion at the bottleneck locations.
3) *Evaluation* – developing and analyzing alternatives for mitigating the congestion at the bottleneck location.
4) *Assessment* – assessing the outcome of the bottleneck removal project after it is implemented in the field.

Guideline 8: Utilize lessons from other successful projects to overcome common barriers

This document is intended to provide individual transportation professionals and the agencies they serve the information needed to overcome common barriers and challenges to implementing low-cost localized congestion relief projects. The lessons learned and unique approaches highlighted in Section 3 should give individuals and agencies confidence that relieving congestion with low-cost improvements is possible and often results in significant operational and safety benefits being realized for relatively little cost.

There is no standard approach to project analysis and development, and other agencies have developed similar frameworks to TTI that are useful for this purpose.

Guideline 9: Agencies should not delay implementation of improvements because most projects pay for themselves in one year

Several case studies, particularly for the 20 projects in Texas, reveal that the benefits of low-cost bottleneck projects typically have benefit-to-cost ratios ranging from 3:1 to 400:1, based on service lives estimated from 5 to 20 years. Further economic analysis reveals that many of these projects actually pay for themselves (i.e., have B/C ratios greater than 1.0) after only one year. This reality supports the guidance that agencies should not delay implementation of improvements because they are extremely cost-effective in an era of constrained funding.

Guideline 10: Skipping the final step (assessment of outcome) takes away the opportunity for lessons learned and potential praise

Several agencies, notably the Minnesota and Texas DOTs, have done an excellent job of making sure to evaluate many of the low-cost localized congestion removal projects they have implemented. Skipping the assessment of outcome of the project leaves the benefits undocumented, removes the opportunity for lessons learned, and also takes away the prospect of potential praise from stakeholders and the media. Taking the time and dedicating the resources up front to make sure assessment of outcome is performed are likely to lead to more successful projects in the future.

Appendix A: Case Study Fact Sheets

Arkansas: Operation Bottleneck Program "Ask & They Will Tell"

Location: Little Rock, Arkansas
Primary Agency: Metroplan
Contact: Jim McKenzie, Executive Director
mckenzie@metroplan.org
(501) 372-3300

Case Study Overview

What was the problem? The federal government requires that metropolitan planning organizations (MPOs) report every 5 years on long-term traffic solutions. The Metroplan MPO determined that federal dollars available for transportation were not meeting local needs and they needed to look into cost-efficient solutions to improving traffic. They also wanted to get a better idea of how bad the regional mobility problems were based on community input.

How was it solved? A comprehensive public outreach effort – *Operation Bottleneck* – was launched in October 2008 to identify traffic bottlenecks as well as auto, bike, and pedestrian safety issues throughout the region. Public input was solicited via newspapers, public meetings, and the internet.

What was the outcome? The *Operation Bottleneck* program received 3,000 responses in 1 ½ months, with on-line submissions constituting the highest return. Metroplan is currently reviewing, classifying, and analyzing the feedback, and the next step will be to coordinate with local jurisdictions. This effort might also help make the case to adopt a Regional Mobility Authority to local agencies that can raise local taxes for public improvements. Other foreseeable activities include: a half-dozen corridor upgrades, an *Operation Roundabout* program to investigate potential intersections, two to three dozen minor spot location projects, hiring a signal timing expert to assist regional agencies with needed adjustments, and consideration of a new traffic management center. One project that has already been implemented is a $1.5 million project for intersection improvements at the Dave Ward Drive/Donaghey Avenue intersection in Conway, Arkansas.

Overcoming Key Challenges & Barriers

What were the barriers & challenges?

1. Disposition towards major projects
2. Lack of confidence in proposed solutions
3. Low-cost solution may blur the need for larger project
4. Lack of understanding of funding possibilities
5. Lack of available resources for implementation

How were they overcome? The mindset of the MPO is now more focused on operations and management and trying to implement solutions that reduce congestion and improve air quality. The region is also pursuing a local tax dedicated to transportation improvements and believes the intense public involvement engendered during the *Operation Bottleneck* program will help to outline specific projects that would be received well by the community.

For more information – http://www.metroplan.org/

I-580/US-101 Connector Ramp Restriping
"It's Amazing What Some White Paint Can Do"

Location: Marin County, California: I-580 WB to US-101 SB
Primary Agency: Caltrans
Contact: Bob Haus, Public Information Officer
robert_haus@dot.ca.gov
(510) 286-5576

Case Study Overview

What was the problem? For years, the one-lane exit ramp between I-580 westbound and US-101 northbound was a daily headache for commuters. First opened in 1956, the connector operated with only one lane for more than half a century, causing one of the most infamous bottlenecks in the Bay Area. It was so notorious that the traffic monitoring company Inrix ranked it as the fourth worst bottleneck in the United States in its 2008 National Traffic Scorecard Report. The company determined that the backed-up exit ramp caused traffic to be congested for 69 hours a week. Cars passed through during peak commute hours at a puttering average of 7.6 miles per hour (mph) due to the abrupt contraction of two lanes to one lane.

How was it solved? In the Spring of 2009, Caltrans – urged by the community – restriped the ramp, temporarily changing the shoulder into a second lane to ease the flow of traffic.

What was the outcome? Positive results from the restriping were almost instantaneous. For 2009, the 580-101 connector ramp plummeted to the inconspicuous position of 491st on Inrix's national bottleneck rankings as drivers experienced only 19 hours of congestion a week and averaged a speed of 14 mph during peak hours.

Overcoming Key Challenges & Barriers

What were the barriers & challenges?

1. Project champion
2. Lack of confidence in proposed solution
3. Low-cost spot solution may blur the need for larger project
4. Lack of shoulders takes away necessary refuge areas

How were they overcome? Caltrans had a significant capital project ($10 million) to improve the ramp to a permanent dual-lane configuration as part of a bigger US-101 project. There was reluctance to implement the project; however, the public basically became the project champion for the interim restriping with their persistent outcry. The lack of shoulders was only going to be a temporary condition because of the already funded project to widen the ramp to two lanes with full shoulders.

For more information – http://bayarea.blogs.nytimes.com/2010/05/21/unclogging-a-traffic-bottleneck-in-marin/

GATEWAY SIGNAL OPTIMIZATION IN BALTIMORE CITY
"REMOVING STOPS MEANS MORE GOS"

Location: Baltimore City, Maryland
Primary Agencies: Baltimore City and Sabra, Wang & Associates
Contact: Ziad Sabra
zsabra@sabra-wang.com
(410) 737-6564

CASE STUDY OVERVIEW

What was the problem? Motorists commuting to and from downtown Baltimore City, Maryland along key arterials faced significant delays at closely-spaced signals. A need was identified to develop optimized signal timing plans to serve both the directional commuter peak as well as local traffic patterns and to reduce travel times, delays, stops and environmental impacts for all users.

How was it solved? A project known as *Baltimore City Gateways Signal Optimization* developed optimized signal timing plans for approximately 175-signals along nine corridors. The City hired a consultant to perform all major components, including data collection, existing conditions evaluation, modeling, timing plan development, field implementation and fine tuning, and travel time studies and recommendations.

What was the outcome? The project had a total cost of $402,500, approximately $2,300 per intersection. The net benefits from retiming the 175 gateway signals alone resulted in significant savings in reduced stops and delays, and fuel consumption; equivalent to a monetary benefit of approximately $20.4 million in the first year (see table below for performance measures).

Benefits of Optimized Gateway Timing Plans

	Vehicle Delay (hr)	Number of Stops	Fuel Consumption (gal)	CO Emissions (kg)	NO Emissions (kg)
"Before"	6,963	499,066	15,827	1,106	215
"After"	4,938	455,807	14,042	982	191
Improvement	2,025	53,259	1,785	125	24
% Improvement	29%	11%	11%	11%	11%
Annual Benefit	$16,221,490	$372,813	$3,213,000	$436,785	$171,936
Total Benefit	$20,416,524				
Approximate Cost	$402,500				
Benefit-Cost Ratio	51:1				

Baltimore City Signal Timing Optimization – Gateway Corridors, May 2008.

OVERCOMING KEY CHALLENGES & BARRIERS

Barriers/Challenges	How They Were Overcome
Lack of training	Hired a consultant with experience in corridor signal timing projects.
Culture of historical practices	Prior to the project the signal shop had a lot of control over signal timings; however majority control was switched to engineering.
No dedicated funding category	City obtained funding using the Congestion, Mitigation and Air Quality (CMAQ) program in addition to local funds.
Lack of available resources	Hiring a consultant supplemented available City staff.

For more information – http://sabra-wang.com/media/BaltimoreSignalTimingOptimizationITEJournalJune08RoikerSilbermanSabra.pdf

GATEWAY SIGNAL OPTIMIZATION IN BALTIMORE CITY

"REMOVING STOPS MEANS MORE GOS"

Location: Baltimore City, Maryland
Primary Agencies: Baltimore City and Sabra, Wang & Associates
Contact: Ziad Sabra
zsabra@sabra-wang.com
(410) 737-6564

CASE STUDY OVERVIEW

What was the problem? Motorists commuting to and from downtown Baltimore City, Maryland along key arterials faced significant delays at closely-spaced signals. A need was identified to develop optimized signal timing plans to serve both the directional commuter peak as well as local traffic patterns and to reduce travel times, delays, stops and environmental impacts for all users.

How was it solved? A project known as *Baltimore City Gateways Signal Optimization* developed optimized signal timing plans for approximately 175-signals along nine corridors. The City hired a consultant to perform all major components, including data collection, existing conditions evaluation, modeling, timing plan development, field implementation and fine tuning, and travel time studies and recommendations.

What was the outcome? The project had a total cost of $402,500, approximately $2,300 per intersection. The net benefits from retiming the 175 gateway signals alone resulted in significant savings in reduced stops and delays, and fuel consumption; equivalent to a monetary benefit of approximately $20.4 million in the first year (see table below for performance measures).

Benefits of Optimized Gateway Timing Plans

	Vehicle Delay (hr)	Number of Stops	Fuel Consumption (gal)	CO Emissions (kg)	NO Emissions (kg)
"Before"	6,963	499,066	15,827	1,106	215
"After"	4,938	455,807	14,042	982	191
Improvement	2,025	53,259	1,785	125	24
% Improvement	29%	11%	11%	11%	11%
Annual Benefit	$16,221,490	$372,813	$3,213,000	$436,785	$171,936
Total Benefit	$20,416,524				
Approximate Cost	$402,500				
Benefit-Cost Ratio	51:1				

Baltimore City Signal Timing Optimization – Gateway Corridors, May 2008.

OVERCOMING KEY CHALLENGES & BARRIERS

Barriers/Challenges	How They Were Overcome
Lack of training	Hired a consultant with experience in corridor signal timing projects.
Culture of historical practices	Prior to the project the signal shop had a lot of control over signal timings; however majority control was switched to engineering.
No dedicated funding category	City obtained funding using the Congestion, Mitigation and Air Quality (CMAQ) program in addition to local funds.
Lack of available resources	Hiring a consultant supplemented available City staff.

For more information – http://sabra-wang.com/media/BaltimoreSignalTimingOptimizationITEJournalJune08RinikerSilbermanSabra.pdf

I-75/M-81 Interchange Reconfiguration
"Michigan Roundabout Proves Golden"

Location: Saginaw County, Michigan: I-75/M-81 interchange
Primary Agency: Michigan Department of Transportation
Contact: Louie Taylor, Delivery Engineer
TaylorLo@michigan.gov
(517) 322-6092

Case Study Overview

What was the problem? Multiple problems existed at this tight diamond interchange. It suffered from heavy truck traffic on M-81, inefficient signal control at the ramp terminals, and safety problems as a result of sharp left turns to and from the ramps that contributed to frequent truck overturns. Backups routinely occurred and the bridge for M-81 over I-75 was in "critical condition."

How was it solved? Given limited resources, the state chose an innovative design approach with roundabouts replacing the tight diamond. The project included reconstruction of the bridge, roundabouts (which include bypass lanes) at the ramp terminals, and removal of the traffic signals. At $5.1 million, this solution represented a cost savings of $6 to $7 million over typical reconstruction, primarily because right-of-way needs were considerably smaller compared to other proposed alternatives.

What was the outcome? The installation of modern roundabouts significantly reduced delay and fuel consumption, with no visible traffic backups. Ultimately, the initially lukewarm reception of the design concept was replaced by public acceptance and accolades for the completed project – including being selected as a Gold Level Winner of the 2007 National Achievement Awards sponsored by the National Partnership for Highway Quality.

Overcoming Key Challenges & Barriers

Barriers/Challenges	How They Were Overcome
Culture of historical practices	Focused discussion with management on design concept – cost savings
Lack of confidence in solution	Simulation models showing level-of-service improvement from D to A.
Getting stakeholder buy-in	Presentations to stakeholders and aggressive public involvement and education campaign including the use of special brochures, videos and newspaper editorials.

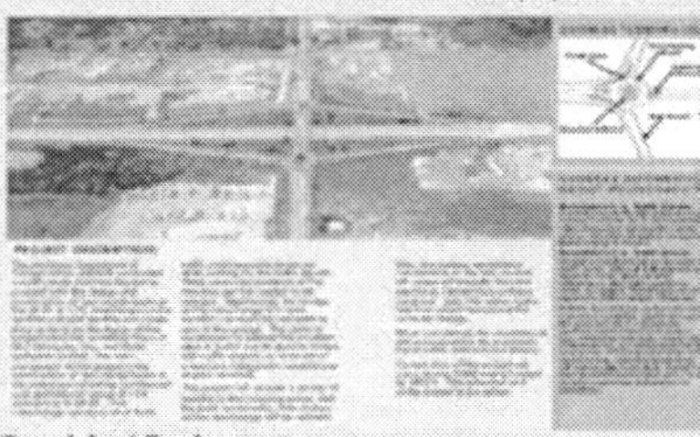

Roundabout Brochure

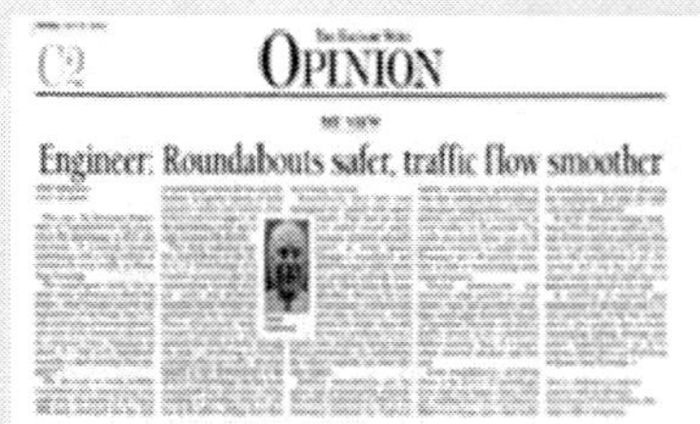

For more information – http://www.nphq.org/doc/awardnominations/2007/Michigan.doc

I-94 LANE MODIFICATION NEAR LOWRY TUNNEL
"PROVIDING OPTIONS CAN SMOOTH FLOW"

Location: Minneapolis, Minnesota: I-94 approaching I-394
Primary Agency: Minnesota Department of Transportation
Contact: John Griffith
John.Griffith@state.mn.us
(651) 234-7728

CASE STUDY OVERVIEW

What was the problem? Congestion regularly occurred in the right lane of westbound I-94 prior to the westbound I-394 exit. This was caused by the high amount of weaving from vehicles that tended to occur from motorists trying to avoid the right lane congestion and merge just north of the Lowry tunnel where there was a lane added.

How was it solved? The $300,000 project reconfigured the westbound lanes so that the right lane through the tunnel became an exit-only lane to I-394, the left lane became exclusively for westbound I-94 and the middle lane was split into two lanes at the tunnel exit with the left lane designated for westbound I-94 and the right lane an "option" lane between either I-94 or I-394. The lane modification was designed to provide better access to I-394, reduce congestion in the right lane, and improve safety in the area by eliminating the last-minute weaving into the right lane.

What was the outcome? Minnesota DOT is gathering data from traffic detectors to regularly assess the performance of this project which was completed in September 2010. Preliminary results show positive increases in throughput during the AM peak and no significant effect in the PM. The safety effects will be evaluated at a later date.

OVERCOMING KEY CHALLENGES & BARRIERS

Barriers/Challenges	How They Were Overcome
Culture of historical practices	Education on benefits in short-term goals.
Lack of confidence in solution	Review of traffic analysis and other data supports the solution.
Spot treatment won't fix the problem	Commitment to monitor for one year and make necessary changes with future programmed project one year out.
Hesitancy to implement non-standard solution	Review of standards and how project does not violate driver expectation and, where design exception needed, determine likely outcomes.

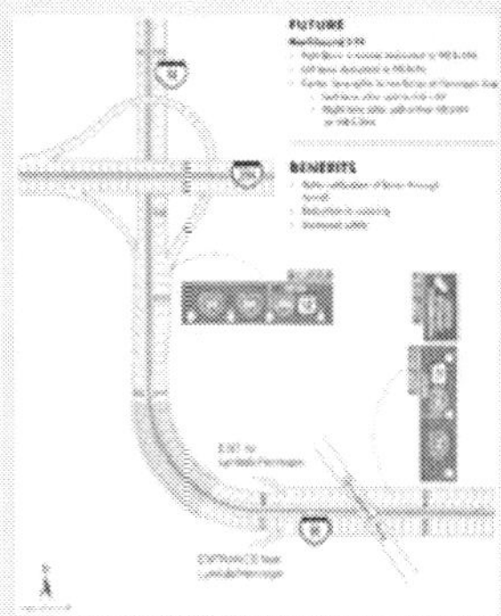

For more information –
http://www.dot.state.mn.us/metro/projects/cmsp/pastprojects.html

I-44/Route 13 Diverging Diamond Interchange
"The Early Bird Gets the Worm"

Location: Springfield, Missouri: I-44 at Route 13
Primary Agency: Missouri Department of Transportation
Contact: Don Saiko, Project Manager
Donald.Saiko@modot.mo.gov
(417) 895-7692

Case Study Overview

What was the problem? The I-44/Route 13 interchange had a number of issues including: traffic congestion, left-turns backing up and blocking thru traffic, and 73% of crashes are rear-end and left-turn right angle. MoDOT had limited funds to improve this location and considered three options: (1) widen the bridge with dual lefts; (2) replace bridge with a Single-Point Urban Interchange (SPUI); or (3) convert the existing interchange to a Diverging Diamond Interchange (DDI), which would be the first in the United States.

How was it solved? MoDOT chose to build the first DDI because it was cheaper to build ($3.2 million vs. $9 million for a SPUI) and maintain. The DDI option also was quicker to build (6 months vs. 1½-2 years) primarily because it utilized the existing bridge and required no additional ROW.

What was the outcome? MoDOT evaluated the DDI on traffic operations, safety & public perception

Traffic Operations	Safety	Public Perception
- Left-turn movements within the DDI had a noticeable decrease in delay & queuing - Through movements within the DDI had a slight increase in travel time - Over-dimension loads up to 18 ft wide and 200 ft long have successful moved - Overall traffic flow is better and typical 1 mile backups during the AM and PM and 2-3 mile backups during weekends and holidays have been eliminated	- Total crashes down by 46% in the first year of operation - Left turn type crashes were eliminated and left turn right angle crashes were down 72% - Rear-end type crashes were down slightly - The DDI's post-construction crash types are similar to any other signalized intersection	- A very high percentage (80+) expressed that traffic flow had improved and delay had decreased - A very high percentage (87%) said that crash were more likely to occur at the previous diamond than the new DDI - A very high percentage (80%) expressed that larger vehicles and pedestrian/bike movements through the DDI were better or similar to a standard diamond

Overcoming Key Challenges & Barriers

Barriers/Challenges	How They Were Overcome
Project champion	The project engineer became champion and garnered by internal (upper MoDOT management) and external (business) support.
Project planning and programming requirements	The FHWA required that a modified Interstate Access Justification (IAJ) be done 2 weeks before the project was let; however, project staff was able to finish the new analysis prior to the deadline.
Lack of confidence in the proposed solution	Project staff utilized an enhanced level of traffic analysis and public outreach (particularly with Wal-Mart) including a simulation model and project video designed to showcase how a DDI operates.
Hesitancy to implement solution that does not follow standards	The project engineer had to convince MoDOT Central Design staff that the pedestrian access would work well through the middle of the DDI instead of on the outside like traditional diamonds.

For more information – http://library.modot.mo.gov/RDT/reports/TRyy1013/or11012.pdf

I-44/ROUTE 13 DIVERGING DIAMOND INTERCHANGE
"THE EARLY BIRD GETS THE WORM"

Location: Springfield, Missouri: I-44 at Route 13
Primary Agency: Missouri Department of Transportation
Contact: Don Saiko, Project Manager
Donald.Saiko@modot.mo.gov
(417) 895-7692

CASE STUDY OVERVIEW

What was the problem? The I-44/Route 13 interchange had a number of issues including: traffic congestion, left-turns backing up and blocking thru traffic, and 73% of crashes are rear-end and left-turn right angle. MoDOT had limited funds to improve this location and considered three options: (1) widen the bridge with dual lefts; (2) replace bridge with a Single-Point Urban Interchange (SPUI); or (3) convert the existing interchange to a Diverging Diamond Interchange (DDI), which would be the first in the United States.

How was it solved? MoDOT chose to build the first DDI because it was cheaper to build ($3.2 million vs. $9 million for a SPUI) and maintain. The DDI option also was quicker to build (6 months vs. 1½-2 years) primarily because it utilized the existing bridge and required no additional ROW.

What was the outcome? MoDOT evaluated the DDI on traffic operations, safety & public perception

Traffic Operations	Safety	Public Perception
– Left-turn movements within the DDI had a noticeable decrease in delay & queuing – Through movements within the DDI had a slight increase in travel time – Over-dimension loads up to 18 ft wide and 200 ft long have successful moved – Overall traffic flow is better and typical 1 mile backups during the AM and PM and 2-3 mile backups during weekends and holidays have been eliminated	– Total crashes down by 46% in the first year of operation – Left turn type crashes were eliminated and left turn right angle crashes were down 72% – Rear-end type crashes were down slightly – The DDI's post-construction crash types are similar to any other signalized intersection	– A very high percentage (80+) expressed that traffic flow had improved and delay had decreased – A very high percentage (87%) said that crash were more likely to occur at the previous diamond than the new DDI – A very high percentage (80%) expressed that larger vehicles and pedestrian/bike movements through the DDI were better or similar to a standard diamond

OVERCOMING KEY CHALLENGES & BARRIERS

Barriers/Challenges	How They Were Overcome
Project champion	The project engineer became champion and garnered by internal (upper MoDOT management) and external (business) support.
Project planning and programming requirements	The FHWA required that a modified Interstate Access Justification (IAJ) be done 2 weeks before the project was let; however, project staff was able to finish the new analysis prior to the deadline.
Lack of confidence in the proposed solution	Project staff utilized an enhanced level of traffic analysis and public outreach (particularly with Wal-Mart) including a simulation model and project video designed to showcase how a DDI operates.
Hesitancy to implement solution that does not follow standards	The project engineer had to convince MoDOT Central Design staff that the pedestrian access would work well through the middle of the DDI instead of on the outside like traditional diamonds.

For more information – http://library.modot.mo.gov/RDT/reports/TRyy1013/or11012.pdf

FREEWAY BOTTLENECK REMOVAL IN TEXAS
"CHAMPIONING LOW-COST PROJECTS"

Location: Dallas-Fort Worth, Texas: 17 projects
Primary Agency: Texas Department of Transportation
Contact: Stan Hall, Advance Planning Engineer
Stan.hall@txdot.gov
(214) 320-6155

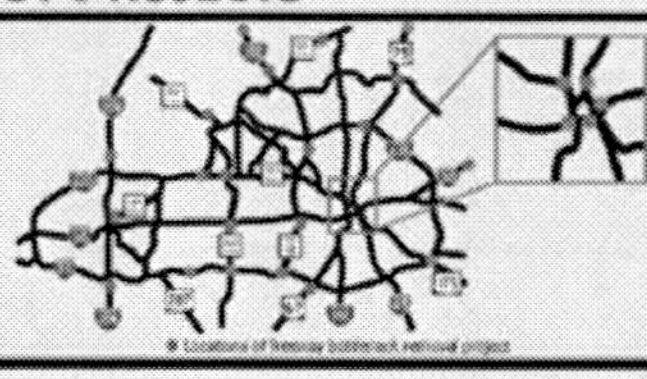

CASE STUDY OVERVIEW

What was the problem?

The DFW area has the most extensive freeway system on a per-capita basis when considering lane-miles and interchanges. Over the last 20 years, recurrent congestion appeared at many locations and in areas with no major capital improvements in the long-range plan. TxDOT developed a number of low-cost projects to address these localized congestion areas, also commonly called bottlenecks. Most improvements were implemented with discretionary funds and project champions were a key to success.

How was it solved?

Some of the typical low-cost solutions implemented in the DFW region included: restriping, reducing lane widths and converting shoulders in order to provide additional travel lane(s) for a short section, auxiliary lanes, and ramp modifications (e.g., closure, reversal and/or metering).

What was the outcome?
See *Transportation Research Record No. 1925*, 2005, pp. 66-75.

Evaluations of the 17 case study projects revealed benefit-cost ratios from 3 to 400:1, based on measured travel time savings. Typical project costs ranged from $5,000 to $2.7 million dollars with most under $1 million. The project also improved safety with an average 35% reduction in injury crash rates after congestion mitigation from the bottleneck improvement. Since TxDOT took the time to evaluate the project outcomes, positive media attention and public feedback from commuters often followed.

OVERCOMING KEY CHALLENGES & BARRIERS

Barriers/Challenges	How They Were Overcome
Project Champion	Support for low-cost bottleneck projects was at a high level, including the District Engineer and MPO Director.
Project planning and programming requirements	Most solutions were over a short distance where air quality and environmental requirements were not necessary.
Lack of training	The Texas Transportation Institute developed a *Freeway Bottleneck* workshop that taught TxDOT about successful bottleneck removal.
Lack of incentives/recognition	TxDOT leadership formalized recognition for successful projects, rewarding Area Engineers during annual performance evaluations.
Design exception process is difficult	Over time, TxDOT gained the trust of FHWA staff by evaluating projects and showing the positive operational and safety benefits.

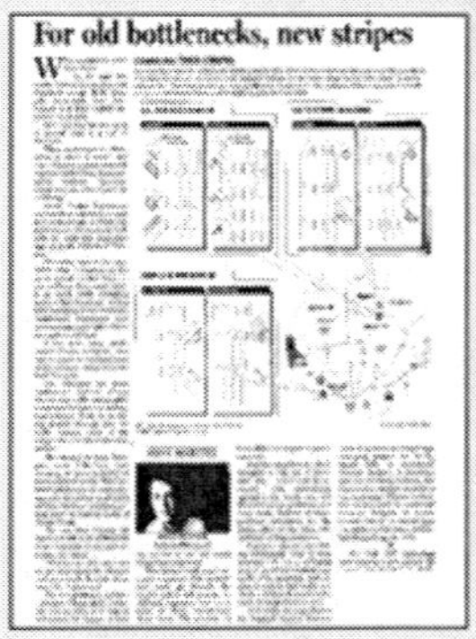
For old bottlenecks, new stripes

APPENDIX B: TABLE 8 INFORMATIONAL LINKS

Listing of Informational Links for Table 8. Overcoming Challenges Matrix (pg 3-3)

Case Studies	Information Link
Little Rock, AR	http://metroplan.org/files/53/3_Metroplan_Unveils_plan_to_battle_traffic.pdf
Phoenix, AZ	http://www.azmag.gov/%5Carchivetotape%5Cpub%5Cbottleneck%5C web site attached file%5Cranking process working paper.pdf
Danbury, CT	http://ops.fhwa.dot.gov/publications/fhwahop09037/examples.htm
Florida DOT Pittsburgh, PA Plano, TX Renton, WA	http://ops.fhwa.dot.gov/publications/fhwahop09037/examples.htm
Kansas City, KS	http://twincitiestransit.blogspot.com/2010/05/kansas-city-next-city-to-use-shoulder.html
LBR workshops	http://ops.fhwa.dot.gov/bn/workshop_series_08/index.htm
Saginaw, MI	http://www.nphq.org/doc/awardnominations/2007/Michigan.doc
Minnesota DOT	http://www.cts.umn.edu/contextsensitive/workshops/flexible/documents/whitepaper.pdf
Manchester, NH	http://www.rebuildingi93.com/
New York, NY Ohio DOT	http://ops.fhwa.dot.gov/publications/fhwahop11009/fhwahop11009.pdf
Rhode Island DOT	http://www.dot.state.
Austin, TX	http://www.dot.state.
Dallas, TX	http://utcm.tamu.edu/publications/final_reports/Walters-Cooner_08-37-16.pdf http://www.lbjexpress.com/ http://www.nctcog.org/trans/cmp/aerial/
Texas DOT Arlington, TX Dallas, TX Fort Worth, TX	http://trb.metapress.com/content/8433746471376934/
Virginia DOT	http://www.dot.state. Forum/HRTPO_Nichols.pdf

In: Traffic Bottlenecks ISBN: 978-1-62081-180-1
Editors: A. L. Sloan and C. F. Dorsey

Chapter 3

AN AGENCY GUIDE ON HOW TO ESTABLISH LOCALIZED CONGESTION MITIGATION PROGRAMS*

United States Department of Transportation
Federal Highway Administration

EXECUTIVE SUMMARY

FHWA's Localized Bottleneck Reduction Initiative (LBR) program has researched over the last five years the causes, impacts, and mitigations available to combat localized recurring congestion; that is, congestion that is primarily "point-specific" as to cause, location, duration, and repetitiveness. This guidance document is intended to aid agencies in establishing either ad hoc or annualized programs that address localized congestion, much in the same way that an annualized safety-spot program would address localized safety issues.

"Localized" recurring congestion differs from regional or corridor-sized congestion, in that the former is characterized by a relatively low-cost, usually correctable operational deficiency (e.g., lane drop, weave, merge, etc.), whereas the latter is often systemic, and may be sufficiently

* This is an edited, reformatted and augmented version of United States Department of Transportation, Federal Highway Administration, FHWA's Bottleneck Reduction Initiative Program Publication, dated March 2011.

complex in cause and relief that funding and/or project duration are typically very high and very long, respectively.

An agency that does not have a stand-alone “localized” congestion program is missing an opportunity to address a specific subset of overall congestion.

> The main questions that this guidance helps an agency frame are:
>
> 1) Do we have a satisfactory agency methodology to specifically address localized congestion problems?
> 2) Within that program, do we have satisfactory justifications for project candidacy, selection, and solutions for said problems?
> 3) Are we executing these projects in a timely fashion and within budgets that are representative of the context of a “Localized Bottleneck Reduction” program?

1.0. Introduction

1.1. Purpose of the Guidance Document

This guidance document provides guidelines that can be used by state DOTs and local transportation agencies in developing a Localized Bottleneck Reduction (LBR) Program. It covers all aspects of designing and implementing an LBR program, from establishing the institutional structure required to support an LBR program to bottleneck identification, improvement, evaluation, and public outreach. The guidance document was developed based on best practices used by state and local agencies.

1.2. How to Use This Document

Target Audiences

This document is designed for state, regional, and local transportation agencies that are focused on mitigating operational causes of bottlenecks. These causes include a wide variety of issues from poorly functioning merges to poor signal progression throughout a corridor. This document targets planners as well as traffic, safety, and design engineers. Operations and Maintenance staff will also find this document as a useful way to begin addressing bottleneck issues in a comprehensive and coordinated way. The document includes a discussion on how best to start a LBR program in your agency. It includes case studies and a series of templates, which the reader can use in starting their own program.

Document Structure and Content

This guidance document includes the following sections:

- **Section 1.0 – Introduction.** This section provides background information on traffic bottlenecks and describes how FHWA is addressing bottlenecks through their LBR Program.
- **Section 2.0 – How to Structure a Localized Bottleneck Reduction Improvement Program.** This section provides guidance to state and local agencies in structuring a program to address localized bottleneck problems. It describes agency roles and responsibilities, options for structuring the program, and factors for success. A self-assessment tool is also provided to assist agencies in establishing where they currently stand with regard to implementing a bottleneck program.
- **Section 3.0 – Resources.** This section provides case study examples of successful bottleneck programs across the United States, including best practice examples in the areas of performance measurement, bottleneck analysis, and project prioritization.

1.3. Background

Bottlenecks: A Definition

The FHWA estimates that 40 percent of all congestion nationwide can be attributed to recurring congestion; some of it "mega" – wherein, entire regions or large facilities (e.g., interchanges or corridors) are overwhelmed by seemingly unceasing traffic demand – and some of it "subordinate" – locations on the highway system where periodic volume surges temporarily overwhelm the physical capacity of the roadway. During off-peak hours, the subordinate locations operate sufficiently and safely for the conditions. These recurring "localized" bottlenecks are those encountered in our everyday commutes, and are characterized as being relatively predictable in cause, location, time of day, and approximate duration. Nonrecurring bottlenecks, on the other hand, are caused by random events such as crashes, inclement weather, and even "planned" events such as work zones and special events.

This guidance document focuses on "localized" recurring bottlenecks (i.e., point-specific or short corridors of congestion due to decision points such as on- and off-ramps, merge areas, weave areas, lane drops, tollbooth areas, and traffic areas); or design constraints such as curves, climbs, underpasses, and narrow or nonexistent shoulders. **Mega-bottlenecks or those occurring due to systemic congestion are not meant to be covered by the scope of this guidance.**

For a bottleneck to be "localized," the factors causing the bottleneck *ideally* should not influence upon, or be influenced by, any other part of the transportation system. As a practical measure, the LBR program recommends considering the closest upstream and downstream decision points as either impacting "to" or impacting "from" the subject location, respectively. Anything much beyond that reach might be considered more than "localized." One exception might be collector-distributor lanes that would almost certainly run through two or more on- or off-ramps. Such a "system" can be considered as a larger, localized condition. Otherwise, recurring, localized bottlenecks generally occur at the following locations:

- A **lane drop**, particularly mid-segment where one or more traffic lanes ends. These typically occur at bridge crossings and in work zones. In large urban areas, a lane drop might occur at jurisdictional boundaries just outside the metropolitan area. Ideally, lane drops should be located at exit ramps where there is a large volume of exiting traffic.
- A **weaving area**, where traffic must merge across one or more lanes to access entry or exit ramps. Bottleneck conditions are worsened by complex or insufficient weaving design.
- **Freeway on-ramps**, where traffic from local streets merge onto a freeway. Bottleneck conditions are worsened on freeway on-ramps without auxiliary lanes, short acceleration ramps, or where there are multiple on-ramps in close proximity.
- **Freeway exit ramps**, which are diverging areas where traffic leaves a freeway. Bottleneck conditions are worsened on freeway exit ramps that have a short ramp length, traffic signal deficiencies at the ramp terminal intersection, or other conditions that may cause ramp queues to back up onto freeway main lanes. Bottlenecks could also occur when a freeway exit ramp shares an auxiliary lane with an upstream on-ramp, particularly when there are large volumes of entering and exiting traffic.

- **Freeway-to-freeway interchanges**, which are special cases on on-ramps where flow from one freeway is directed to another. These are typically the most severe form of physical bottlenecks because of the high traffic volumes involved.
- **Abrupt changes in highway alignment**, which occur at sharp curves and hills and cause drivers to slow down either because of safety concerns or because their vehicles cannot maintain speed on upgrades. Another example of this type of bottleneck is in work zones where lanes may be redirected or even slightly shifted during construction.
- **Low clearance structures, such as tunnels and underpasses.** Drivers slow to use extra caution, or overload bypass routes. Even sufficiently tall clearances could cause bottlenecks if optical illusion causes a structure to appear lower than it really is, causing drivers to slow down.
- **Lane narrowing**, caused by either narrow travel lanes or narrow or nonexistent shoulders.
- **Intended interruptions to traffic flow** that are necessary to manage overall system operations. Traffic signals, freeway ramp meters, and tollbooths can all contribute to disruptions in traffic flow.

A detailed discussion on bottleneck characteristics is provided in FHWA Publication FHWA-HOP-09-037, *Recurring Traffic Bottlenecks: A Primer – Focus on Low-Cost Operational Improvements*, available on FHWA's web site at http://www.ops.fhwa.dot.gov/bn/index.htm.

Bottlenecks: A History

Timeline of National Bottleneck Activities

Over the past several years, transportation professionals have come to realize that highway bottlenecks demand special attention. Several national studies have highlighted bottlenecks as a major congestion problem in urban areas. These studies have raised the level of awareness about bottlenecks as a problem, warranting that they be treated as a significant part of the congestion problem.

One of the LBR tenets is "a bottleneck is congestion, but congestion is not always just a bottleneck." This means that a bottleneck (or chokepoint) is merely a subset of the larger congestion pie.

However, that "subset" is now realized to be a uniquely impacting (and increasingly growing) genre of congestion; namely, that it is *subordinate* locations along a highway that need to be fixed, and not necessarily the knee-jerk expectation to rebuild the entire facility.

Granted, in some cases, an aging or clearly capacity-deficient facility may need to be replaced. But in this age of budget constraints and economizing, one or two corrections to inefficient subordinate locations on a facility may be all that is needed to improve the condition.

The American Highway Users Alliance (AHUA) conducted studies of the nation's urban bottlenecks in 1999 and 2004. The studies produced rankings of the worst bottlenecks in terms of total delay to travelers and discussed what was being done to fix the problems at locations where specific improvements had been scheduled.

The studies found that nearly all of the *worst* bottlenecks are major freeway-to-freeway interchanges in large urban areas, but many smaller bottlenecks were surprisingly impacting as well.

The 2004 study updated the 1999 rankings and discussed three bottleneck improvement success stories – bottlenecks identified in 1999 that were subsequently improved or well under construction; the clear message being that mitigations can be realized.

States and regions are beginning to recognize the significance of bottlenecks as well. The Ohio Department of Transportation completed a study of freight (trucking) bottlenecks, and the Interstate 95 Corridor Coalition is undertaking a study of all potential bottlenecks in Coalition states. The Atlanta Regional Commission has defined bottlenecks as a specific portion of their Congestion Management Process and is identifying regional and local bottlenecks in their network.

In 2002, the Texas Transportation Institute compiled a database of before/after measurements of selected Texas DOT initiatives to remove bottlenecks and improve operations on urban freeways. The benefit/cost ratios and reductions in crash rates observed at these locations provided valuable insight regarding the effectiveness of various bottleneck reduction strategies over time.

More recently, an effort by a private data provider, Inrix, also identified the nation's worst bottlenecks (Table 1.1). Whereas previous bottleneck identification efforts were based on analytic procedures using traffic volumes and capacity data, Inrix's approach uses data assembled by them from a variety of sources. As direct travel time measurements become more common and better refined, the science of bottleneck identification and performance will improve.

Table 1.1. Inrix 2009 Annual Report Worst Traffic Bottlenecks

Rank									
2009	2008	CBSA (Pop Rank)	Road/Direction	Segment/Interchange	County	State	Length (Miles)	Hours of Congestion	Average Speed While Congested
1	1	New York (1)	Cross Bronx Expy WB/ I-95 SB	Bronx River Pkwy/ Exit 4B	Bronx	NY	0.35	94	11.4
2	7	Chicago (3)	Dan Ryan Expy/ I-90/I-94 WB	Canalport Avenue/ Cermak Road/Exit 53	Cook	IL	0.4	83	11.1
3	5	New York (1)	Cross Bronx Expy WB/ I-95 SB	I-895/Sheridan Expy/ Exit 4A	Bronx	NY	0.51	95	12
4	2	New York (1)	Cross Bronx Expy WB/ I-95 SB	White Plains Road/ Exit 5	Bronx	NY	0.28	86	11.9
5	3	New York (1)	Harlem River Drive SB	3rd Avenue	New York	NY	0.16	77	9.6
6	62	New Haven (58)	I-91 SB	Hamilton Street/ Exit 2	New Haven	CT	0.22	76	12.8
7	13	Los Angeles (2)	Hollywood Fwy/ U.S. 101 NB	Los Angeles Street	Los Angeles	CA	0.1	85	14
8	24	Chicago (3)	Dan Ryan Expy/ I-90/I-94 WB	18th Street/ Exit 52C	Cook	IL	0.41	83	13.7
9	11	New York (1)	Cross Bronx Expy WB/ I-95 SB	Westchester Avenue/ Exit 5	Bronx	NY	1.15	76	12.8
10	26	Chicago (3)	Dan Ryan Expy/ I-90/I-94 WB	Ruble Street/ Exit 52B	Cook	IL	0.12	86	14.7
11	10	Los Angeles (2)	Hollywood Fwy/ U.S. 101 SB	Vermont Avenue	Los Angeles	CA	0.62	82	15.4
12	16	Los Angeles (2)	Hollywood Fwy/ U.S. 101 NB	Alameda Street	Los Angeles	CA	0.27	82	14
13	29	Chicago (3)	Dan Ryan Expy/ I-90/I-94 WB	Roosevelt Road	Cook	IL	0.22	86	16.2

Table 1.1. (Continued)

Rank		CBSA (Pop Rank)	Road/Direction	Segment/Interchange	County	State	Length (Miles)	Hours of Congestion	Average Speed While Congested
2009	2008								
14	9	New York (1)	Harlem River Drive SB	2nd Avenue/125th Street/ Exit 19	New York	NY	0.22	75	11.1
15	6	New York (1)	Van Wyck Expy/I-678 NB	Liberty Avenue/Exit 4	Queens	NY	0.52	61	11.7
16	36	Los Angeles (2)	Pasadena Fwy/ CA 110 NB	Sunset Boulevard/ Exit 24A	Los Angeles	CA	0.21	45	10.4
17	12	New York (1)	Van Wyck Expy/I-678 NB	Hillside Avenue/Exit 6	Queens	NY	0.12	71	14
18	8	Chicago (3)	Eisenhower Expy/ I-290 EB	U.S. 20/U.S. 45/ U.S. 12/Exit 17	Cook	IL	0.99	62	14.3
19	19	Los Angeles (2)	Hollywood Fwy/ U.S. 101 NB	Spring Street	Los Angeles	CA	0.14	82	16.1
20	17	New York (1)	I-95 NB	U.S. 9/U.S. 1/ U.S. 46/Exit 72	Bergen	NJ	0.42	66	11.3

FHWA Involvement

FHWA's first effort related to bottlenecks was in the freight (trucking) arena. In 2005, using the AHUA studies as a starting point, FHWA conducted an assessment of the impacts that highway bottlenecks have on truck freight shipments. Bottlenecks outside of urban areas were also considered (e.g., steep grades). **A major finding of this study was that in terms of total delay, the urban bottlenecks – typically thought of as a commuter-related problem – are also major sources of truck delay.** In 2009, in partnership with the American Transportation Research Institute (ATRI), FHWA conducted an in-depth analysis to produce a congestion severity ranking for 100 freight significant highway interchanges. The rankings can be used by both public and private sector stakeholders to better identify transportation system deficiencies and to assist in logistics and routing decisions. ATRI and FHWA will continue to monitor these locations on an annual basis.

In preparing the FY 2006 Strategic Implementation Plan (SIP), the FHWA Operations Leadership Council discussed the need for a national strategy to address bottlenecks and requested each Division office to identify bottleneck locations in their state. The responses ranged from mega bottlenecks encompassing multi-interchange corridors to point-specific localized chokepoints. FHWA initiated the Localized Bottleneck Reduction (LBR) Program in 2007 to raise awareness of point-specific localized bottlenecks at the state level and promote low-cost, quick-to-implement geometric and operational improvements to address recurring chokepoints. One of the first activities of the LBR Program was a survey of FHWA Division office personnel to obtain state best practices in bottleneck identification, assessment, countermeasures, and evaluation, including examples of overcoming unique challenges such as strategies requiring a design exception, cooperation among several departments, creative funding, and public perceptions. A compendium of state best practices was developed from these responses. More details on the LBR Program are provided in Section 1.4.

Previous Similar Bottleneck Efforts

Topics

The Federal-Aid Highway Act of 1968 established the Traffic Operations Program to Improve Capacity and Safety (TOPICS). The program authorized $200M in Federal matching to fund projects designed to reduce traffic congestion and facilitate the flow of traffic in urban areas. Although candidate projects were not planned to the same scale as regional planning processes,[1]

TOPICS was a landmark program in demonstrating the concepts and effectiveness of traffic management practices.

1.4. The Localized Bottleneck Reduction Program

FHWA's Localized Bottleneck Reduction (LBR) Program promotes operational and low-cost bottleneck mitigation strategies to improve mobility. Managed by the Office of Operations, the program serves to bring attention to the root causes, impacts, and potential solutions to traffic chokepoints that are recurring events; ones that are wholly the result of operational influences. This is "good and bad" news in the sense that design influences can always be corrected, but some corrections may be cost-prohibitive in terms of direct construction costs or indirect right-of-way impacts. Regardless, a significant amount of locations can be corrected for relatively low-cost and/or low physical impact. In any case, the goal of the program is to raise awareness of bottlenecks at the state level and promote low-cost, quick-to-implement geometric and operational improvements to address recurring chokepoints. The LBR Program has several activities either completed or underway, including:

- This guidance document, which provides guiding principles and concepts common to low-cost operational improvement programs to assist states in adopting a programmatic approach to addressing traffic congestion at bottlenecks.
- *Traffic Analysis Toolbox Volume X: Localized Bottleneck Congestion Analysis – Focusing on What Analysis Tools are Available, Necessary, and Productive for Localized Congestion Remediation.* This document provides guidance on analysis tools and data inputs required to analyze the specific genre of localized congestion problems.
- *Recurring Traffic Bottlenecks: A Primer – Focus on Low-Cost Operational Improvements.* This Primer is the "face" of the program. It provides an overview of the wide range of operational and low-cost strategies available to reduce congestion at bottlenecks.
- A compendium of state best practices in bottleneck identification, assessment, countermeasures, and evaluation, including how

bottlenecks are treated in the annual planning and programming processes.

- Localized Bottleneck Reduction Regional Workshops. Regional workshops for state and local agencies to learn and share information on localized bottleneck reduction strategies and how they can be incorporated into state and local planning processes.
- Many of the items listed above can be found at the FHWA bottleneck web site, which can be found at the FHWA Office of Operations web site.

Additional guidance documents are forthcoming that are aimed at agencies and personnel who have first responsibility to address bottleneck congestion locations.

2.0. How to Structure a Localized Bottleneck Reduction Improvement Program

2.1. Roles and Responsibilities

There really are no set guidelines for roles and responsibilities of an LBR program. State DOTs, MPOs, or local transportation agencies could all lead an effective LBR effort. State DOTs and MPOs are traditionally the organizations who lead LBR efforts simply because they usually have larger missions, which include bottleneck issues; as well as access to a variety of funding mechanisms. Many successful LBR programs actually depend on a high level of coordination between state DOTs and MPO. Many times, the state may identify bottlenecks and work closely with MPOs to integrate these projects into their TIP and other targeted funding sources such as CMAQ and safety. However the split or leadership role, any transportation agency can lead an effective program.

2.2. Options for Structuring the LBR Program

Those agencies that have been effective at dealing with bottlenecks have developed either special or ongoing programs specifically targeted at dealing with current bottleneck projects. The options for structuring an LBR program vary widely, as described in this section.

Barriers to Establishing the LBR Program

States have cited a number of barriers to establishing bottleneck-specific or similar programs that target chokepoint congestion:

- **A predisposition for large scale, long-term congestion mitigation projects**. Traditional transportation planning and programming efforts are often predisposed toward major capital improvement projects to relieve congestion such as corridor-widening or massive reconstruction of an interchange. There is also no shortage of demand management strategies designed to fight the congestion battle, such as HOVs, tolling and pricing, transit alternatives, and ridesharing programs. But the onerous processes involved in many of these initiatives can squeeze out smaller programs.
- **Lack of program identity**. Unless there is a formal program identity, bottleneck remediation is usually relegated to a few projects completed as part of an annualized safety program, or as a subordinate part of larger, other purposed projects.
- **Lack of a champion**. Many successful state or metropolitan planning organization programs are the result of one or more persons taking charge to either mandate or adopt a program. High-level administrators often set the policy direction and strategic initiatives for their agencies, while mid-level managers' production reflects their priorities and skills in executing those initiatives.
- **Lack of resources**. Many state agencies are finding themselves overworked and understaffed. Although the return on investment for LBR projects are high, agencies often do not have the in-house resources necessary to conduct detailed analyses required to evaluate and prioritize the large number of potentially competing projects. With limited resources, agencies are relegated to hiring consultants and/or universities to conduct detailed project analysis.
- **Lack of funding**. With many state agencies experiencing major budget shortfalls, lack of funding continues to be an often cited barrier to implementing new programs.
- **Responsibility has not been assigned.** Not part of ongoing planning and programming processes. Localized bottleneck mitigation projects are not often included in the ongoing planning and programming

processes for most agencies. Others struggle with how best to identify problem locations, assess existing conditions, and quantify the impacts of proposed remedies, as there is no structured process in place. For example, in developing their structured LBR program, Michigan DOT cited challenges regarding how best to justify and evaluate project impacts while creating a level playing field for application of LBR funding across each of their seven regions.

- **A culture of historical practices**. Many agencies face institutional challenges in changing their current business practices. For example, one agency dutifully executed an annualized "safety" program and looked only at crash rates in determining their annual top 10 list of projects. After instituting a congestion mapping process, they identified several significant stand-alone chokepoints that did not correlate with their high-crash mapping. Thereafter, high congestion hot spots competed with high accident hot spots on their unified top 10 list of projects.

The options for structuring an LBR program vary widely. One approach is to undertake a one-time special program or periodic special initiatives that focus on bottlenecks. Other approaches incorporate annualized reviews or ongoing programs – similar to annualized safety-spot lists – that organized candidate locations and identify them for relatively short-term (e.g., two years or less) attention.

Periodic Special Program or Initiative

For example, in 2007, **the Minnesota DOT** was asked by the Legislature to develop a rapid turnaround plan to identify low-cost, quickly implementable **projects that were *not already identified* by the traditional planning and programming processes.** In a matter of months, this unique approach led by the Traffic Management Center engineers basically "brainstormed" low-cost, candidate projects that were nagging problems, but for whatever reasons, had never landed on traditional Capital Improvement Programs. In 2008, the **Central Arkansas MPO** undertook "Operation Bottleneck," a campaign to openly solicit public input of candidate locations, but one that has a finite life span.

Due to resource limitations, some agencies are using research funds to sponsor special projects involving in-depth analysis of the causes and temporal variations at key bottleneck locations. For example, in the Phoenix metropolitan region, **Arizona DOT** commissioned a special research study to identify freeway bottlenecks and provide a detailed assessment and recommendations of countermeasures. **Wisconsin DOT** partnered with a research agency to identify bottleneck reduction strategies and to develop a *Paramics* model to perform in-depth simulation of the proposed strategies. They plan to expand the project to include other bottleneck locations statewide.

Incorporating Bottlenecks into Other Programs

State DOT Level

At the state DOT level, low-cost bottlenecks can be addressed programmatically even without a special program or initiative. One approach is to conduct a review of existing plans and look for opportunities to include LBR improvements in them. For example, **Caltrans**, as part of their Corridor Management Process, includes the identification of bottlenecks and potential short-term fixes as part of an overall and long-term strategy for making corridor improvements. This often takes the form of an "LBR audit," which is a review of traditional large-scale corridor studies to identify opportunities for using LBR improvements as part of the package of improvements. The concept here is similar to that of Road Safety Audits.

Another approach is to integrate bottleneck strategies as part of ongoing planning and processes, thereby incorporating them into an agency's congestion mitigation toolbox. For example, the **Ohio DOT** added a congestion-based index ranking to their annual identification of spot safety problems for the Federal Hazard Elimination Program (HEP). As a result, congestion hot spots now have a "voice" regardless of crash indices, and congestion-related projects compete for attention in the project selection and scheduling process.

Washington State DOT recognizes bottlenecks and chokepoints as an integral part of their project planning and development process. The recent *Moving Washington* initiative incorporates LBR concepts into a coordinated program to address congestion. In the *Moving Washington* program, WSDOT applies three balanced strategies to fight congestion – operate efficiently, manage demand, and add capacity strategically. By strategically adding capacity, WSDOT targets bottlenecks and chokepoints in the transportation system, and does so cost-efficiently. Recent funding shortfalls have made low-cost, strategic investments a high priority with WSDOT and LBR projects fill this role nicely. Performance results show that *Moving Washington* strategies and projects are making a difference around the State to relieve congestion.

At the planning stage in their Highway System Plan, WSDOT considers bottlenecks together with traditional corridor improvements under the "Congestion Relief" category. Congestion relief projects are ranked using the benefit/cost ratio, contribution to performance goals, and other qualitative factors, and compete on these bases with projects in other categories in the Highway System Plan: Preservation, Safety, Environmental Retrofit, Economic Vitality, and Stewardship.

MPO Level

At the metropolitan planning organization level, the short-term nature of LBR projects meshes well with the Congestion Management Process (CMP) and "planning for operations," which are new initiative areas for planners. As planners' perspectives broaden to include these short-term views of the system (in addition to the traditional long-range view), an LBR program makes perfect sense. In fact, one idea may be to include consideration of an LBR program within one the CMP and/or "planning for operations" guidance being developed by FHWA. The idea is not to make an LBR program subordinate to these initiatives but rather to integrate it into the initiatives, noting that LBR is an effective way to address current congestion problems.

From a planner's viewpoint, LBR improvements would be another aspect of the CMP process. Because an LBR program should be data- and performance-driven, it is a logical complement to a CMP; the same data should be used for both purposes. In fact, the steps required to implement an LBR program directly mimic the eight-step CMP process: problems are identified, countermeasures are developed to address the problems, funding is identified, and projects are implemented, and evaluation occurs. Figure 2.1 shows the eight-step CMP process, which bears remarkable similarity to the steps that need to occur within an LBR program. An LBR program is a natural extension of what planners are already doing. In fact, within the context of the CMP, it may useful to make the two processes seamless, at least at the MPO level.

Therefore, there is a strong case for encouraging planners to be engaged in the development and operation of an LBR program.

Identified LBR Program

Another option is to establish a defined bottleneck program within the agency. For example, **Virginia DOT** has implemented the Strategically Targeted Affordable Roadway Solutions (STARS) Program, which is a safety and congestion program that partners state, planning district and local transportation planners, traffic engineers, safety engineers and operations staff to identify "hot spots" along roadways where safety and congestion problems overlap and are suitable for short-term operational improvements. In summary, the STARS program "road audits" are shelf-ready when budget and/or opportunity (e.g., proximity development plans) are available.

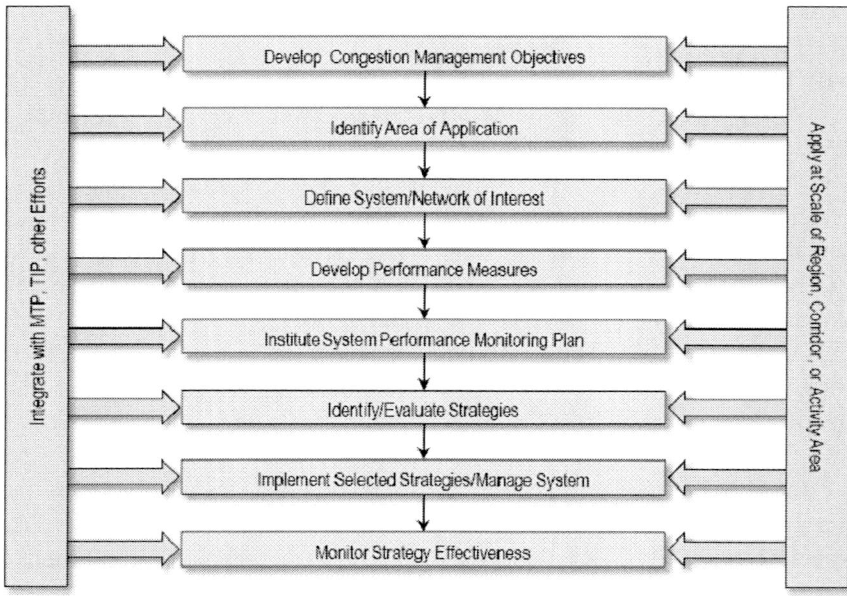

Figure 2.1. Steps in the Congestion Management Process.

Some agencies were originally driven to explore low-cost congestion relief projects because of budgetary restrictions, but soon realized that these projects could be implemented very quickly and, as a bonus, were highly visible and popular with the public. Several DOTs have been able to create a defined bottleneck program area as a result of their success with special initiatives such as these. For example, as a result of their initial success, **Minnesota DOT** was able to develop a highly accelerated process for bottleneck identification and prioritization, which led to many effective projects in the following two years. Mn/DOT also found that because of lower costs, it could identify multiple locations throughout the region and "spread around" bottleneck reduction projects in a fair and equitable manner. **Utah DOT** initiated a Choke Point Program in 2006 to address safety, congestion, and bottleneck areas. It was so successful that the Utah Legislature put forth funding for additional chokepoint projects.

Other agencies are developing dedicated LBR Programs in response to FHWA Localized Bottleneck Reduction Workshops held around the United States in 2009 and 2010. These include **Indiana DOT** and **New York DOT**.

2.3. Factors for Success: What Does an LBR Program Have to Consider?

The institutional or policy component of localized bottlenecks is extremely important because of the way that agencies and processes have been structured.

This is similar to the situation that systems operations and management (O&M) faced 10 to 15 years ago: there was no organizational unit set up to deal with operations and management. As a result, O&M struggled to find an identity in most agencies, and development of O&M programs were hindered. Likewise, LBR programs face institutional and policy barriers.

Therefore, one key aspect is to understand the challenges facing agencies in defining a programmatic structure for dealing with bottlenecks. Some of these issues include:

- Competition for funding with other traditional programs;
- Implementation issues such as meeting design standards and conformity requirements;
- Choosing between a temporary, permanent, or periodic (recurring every few years) program for an LBR program;
- Creating a new organization versus imbedding an LBR program within a current program area (e.g., planning or maintenance); and
- Developing management support for an LBR.

This section presents success factors for overcoming these challenges in order to implement a LBR program.

Setting Goals and Objectives for an LBR Program

Setting goals and objectives is an important part of framing the establishment of an LBR Program within an agency. The following success factors were identified from case study examples:

- Relate the goals of the LBR program to the operating objectives of the agency as a whole;
- Gain the public's support in targeting and fixing smaller bottleneck issues in a timely manner;
- Utilize rapid methods to identify and evaluate bottlenecks; and
- Create separate funding mechanism for bottleneck mitigation or ensure bottleneck projects can compete equitably through traditional methods.

New York DOT's Region 11 Planning Office (New York City) established their LBR Program with a mission to mitigate recurring congestion at selected chokepoints along the region's highways, local streets, and intersections, and the operational influences that cause them. Their goal was to investigate the opportunities as well as develop measures for the application of operational and low-cost infrastructure improvements to address chokepoints and to identify cost-effective improvements either as stand-alone initiatives or as part of existing Capital Projects, with an overall objective of moving project recommendations into the capital program.

Championing an LBR Program

The need and urgency for establishing a LBR Program is not always shared across all levels of an agency. Nevertheless, a clear case must be established with upper management and elected officials to secure resources and commitment to proceed with LBR activities.

One way to accomplish this is to **demonstrate the return on investment** to the organization regarding the implementation of an LBR Program, as well as developing a sustainable approach for assessing the impacts of LBR improvements.

For example, in developing their structured Bottleneck Reduction Program, Michigan DOT's Systems Operations and Management (SOM) Section sought to obtain leadership support by requesting a dedicated funding template specifically for bottleneck reduction projects. MDOT does not have any dedicated funding at this time, but it is being discussed. They have developed a systematic approach for demonstrating excellent benefit-to-cost ratios among Bottleneck Reduction projects, which helped to justify the creation of the statewide Bottleneck Reduction Program. MDOT is performing analysis on existing road improvement projects so recommendations can be implemented within the existing projects until dedicated funding is approved.

Other ways to raise visibility include:

- Appointing one element (person or office) within an organization that is the **single point of contact** and champion for the LBR program. Having a sustained champion can greatly enhance the effectiveness of the program within an organization.
- **Involving the public** in identifying bottlenecks is another way to raise the visibility of the program while at the same time building momentum within the organization.

Personnel/Unit Responsible for the LBR Program

The roles and responsibilities for planning and operations usually fall within separate and distinct departments within a typical state DOT. Coordination between departments can improve how an agency tackles bottleneck congestion. A review of state DOTs that have successfully implemented LBR programs within their organizations revealed that having an interdisciplinary team or committee responsible for administering the LBR program is a useful consensus building tool. These teams would facilitate activities of mutual interest such as:

- Identifying strategies to address bottlenecks;
- Developing bottleneck performance measures that are consistent across the agency;
- Coordinating data collection and developing tools for data analysis;
- Developing models to quantify the impacts of bottleneck strategies; and
- Overseeing implementation of bottleneck mitigation strategies and objectives through the TIP or LRTP.

For example, **New York DOT's Region 11** LBR Program is administered through an interdisciplinary project scoping team headed by the Planning Director, but whose members are comprised of staff from other functional groups within the agency. Such an approach is a useful tool for building consensus on the nature of transportation problems, the key issues, and best strategies to address bottlenecks. It also creates a forum for early buy-in of preferred alternatives and courses of action across each of the various functional groups.

Indiana DOT is establishing their LBR Program as part of an overall organizational change as they transition to a new funding system that is asset management based with multiple funding teams. As part of the new structure, they are establishing a Mobility Asset Team.

LBR Program Funding

This section presents success factors and examples of how states/MPOs have funded LBR projects. Options include:

- Developing a funding decision matrix where bottleneck countermeasures compete for funds alongside other mobility and congestion management projects based on benefit-to-cost ratio;
- Funding low-cost improvement projects as part of a dedicated signal timing program; and
- Implementing a dedicated program to provides funding for districts/regions to implement low-cost improvement projects.

For example, the Mobility Asset Team (described shortly above) at **Indiana DOT** will fall under the LBR funding category. INDOT expects that bottleneck countermeasures will have a very good benefit-to-cost ratio and projects will be competitive in the funding decision matrix that is part of the new funding system.

In 2008, **Ohio DOT** initiated a new Systematic Signal Timing Program that provides district offices and local governments with technical assistance (task order consultants) to analyze, coordinate, and upgrade the timing and phasing of signal systems in high-crash areas. Funding can also be used to upgrade signal equipment as needed. Funding is targeted to those areas listed as a top crash priority by ODOT, MPOs, or local governments. Project sponsors must verify that the crash patterns are relevant to signal timing concerns. Funding requests are funneled through the appropriate ODOT district office, which typically must concur with the request for assistance.

Funded by the Utah Legislature, the **Utah Department of Transportation** (UDOT) initiated a Choke Point Program in 2006 to address safety, congestion, and bottleneck areas. It was successful, so the Legislature put forth more funding for additional chokepoint projects. The UDOT has used the Choke Point Program to identify potential projects that could improve operations and safety with relatively low cost and in a short time period.

Pennsylvania has in place a Congestion Corridor Improvement Plan (CCIP) which provides funding for districts/regions to study low-cost improvements. However, this program has not been funded since 2008. It is hoped that the program will be resumed whenever funding is available.

The LBR Planning Process

Planning for localized bottleneck improvements includes a number of steps that could happen in a number of ways. But regardless of the approach, they can be loosely organized into the following five broad steps:

1) Bottleneck Identification;
2) Identify Bottleneck Improvement Strategies;
3) Prioritize Projects;
4) Programming and Implementation; and
5) Evaluate Bottleneck Improvement Projects.

The remainder of this section expands on each of these steps. Later in Section 2.0 of this document, the unique approaches an agency could take in starting their own LRB program are presented but they are all variations on these steps.

Bottleneck Identification

There is a variety of methods for bottleneck identification. One of the primary methods is a data-driven approach, where roadway characteristics data are used to identify chokepoint locations. Potential data sources include ITS data from traffic management centers, or roadway characteristics and traffic data from statewide traffic and HPMS monitoring locations. Traffic analysis tools can be used to systematically identify bottleneck locations by analyzing road segments for congestion or poor levels of service. For example, at **Caltrans**, system monitoring and evaluation using the Performance Monitoring System (PeMS) is seen as the foundation for the entire Corridor System Management Plan process because it cannot only identify congestion problems, but can also be used to evaluate and prioritize competing investments.

Bottleneck locations can also be identified anecdotally by tapping into local agency knowledge and personal experience (e.g., district-level), direct observation of local agency personnel, aerial photographs, or video surveillance data. For example, when **Michigan DOT** solicited potential bottleneck locations, they were sure to include problem descriptions from each of their seven region offices and not merely their headquarters' analysis. More than 200 locations were identified, with about one-third being freeway interchanges.

There is a hybrid approach where anecdotal information and judgment is used to identify candidate locations but data are used to assess priorities objectively. For instance, the following conditions typically exist or help to identify a recurring bottleneck condition:

- A traffic queue upstream of the bottleneck, wherein speeds are below free-flow conditions elsewhere on the facility.
- A beginning point for a queue. There should be a definable point that separates upstream and downstream conditions. The geometry of that point is often coincidently the root cause of the operational deficiency.
- Free flow traffic conditions downstream of the bottleneck that have returned to nominal or design conditions.
- As it pertains to an operational deficiency, a predictable recurring cause.
- Traffic volumes that exceed the capacity of the confluence to process traffic. (Note: This applies to recurring events even more so than nonrecurring.)

Some metropolitan areas have even conducted public outreach efforts to solicit input from motorists on bottleneck locations. For example, **Metroplan, the MPO for the Little Rock, Arkansas region**, conducted Operation Bottleneck during the fall of 2008, a public outreach effort designed to identify traffic bottlenecks, as well as automobile, bicycle, and pedestrian safety issues throughout the region. The program received 3,000 responses within two months, with on-line submissions constituting the highest return. Several minor roadway improvements have already been completed or are planned as a result of the program.

Bottleneck Improvement Strategies

This section describes the types of LBR treatments, as well as a matrix on matching bottleneck treatments to needs. It also describes various barriers to implementing LBR treatments such as design standards/exceptions, air quality conformity, and consistency with long-range design concepts.

Types of LBR Treatments

The following is a sampling of short-term, low-cost operational and geometric improvements. All of these remedies address operational deficiencies, as opposed to other congestion mitigation efforts that address driver choice, travel demand, corridor-wide upgrades, or simply (but expensively) building our way out of congestion.

1) **Shoulder conversions.** This involves using a short section of traffic bearing shoulder as an additional travel lane. Shoulder conversions are appropriate between interchanges or to provide lane congruency with adjacent sections. The shoulder condition should be rated for use as a travel lane.
2) **Restriping merge or diverge areas** to provide additional lanes, provide an acceleration/deceleration lane, extend the merge/diverge area, or improve geometrics to better serve demand.
3) **Lane width reductions.** This involves reducing lane widths and restriping to add an additional travel and/or auxiliary lane.
4) **Modify weaving areas** by adding collector/distributor or through lanes.
5) **Ramp modifications.** These could include ramp metering; widening, extending, closing, or consolidating ramps; or reversing entrance and exit ramps to improve operations.
6) **Speed harmonization (variable speed limits)**. This is the practice of adjusting speed limits when congestion thresholds have been exceeded and congestion and queue forming is imminent. Speed harmonization can also be used to promote safer driving during inclement weather conditions. This mostly European practice reduces the traffic "shock wave" that results through congested corridors, thereby delaying the onset of a breakdown in traffic conditions. The result is decreased headways and more uniform driver behavior, which indirectly benefit bottlenecks and chokepoints.
7) **"Zippering" or self-metering that promotes fair and smooth merges.** A motorist who is 10th in line knows that he will be 20th to merge into the single lane ahead. This helps to eliminate line jumpers that bull ahead, disrupt the queues, and often block adjacent lanes until they force their way in line. Usually this method of merging requires on-site enforcement, but often is exhibited by regulars who know the process and are willing to abide.
8) **Improve traffic signal timing on arterials.** Also, traffic signal timing improvements at ramp terminal intersections will prevent ramp queues from backing up onto freeway main lanes.
9) **Access management principles** to reduce vehicular conflicts (hence, delays) on arterial corridors
10) **Continuous flow intersections.** These are unconventional at-grade intersections, which eliminate one or more left-turn conflicts at a main intersection. This is achieved through dedicated left-turn bays located

several hundred feet prior to the main intersection, which allow left-turning vehicles to move at the same time as through traffic. The left-turn traffic signal phase is eliminated, allowing more vehicles to move through the main intersection and thus reducing traffic congestion and delays. These at-grade intersections achieve traffic flow similar to grade-separated interchanges, but at a considerably lower cost.

11) **High-Occupancy Vehicle (HOV) or reversible lanes.**
12) Provide **traveler information** on traffic diversions.
13) **Implement congestion pricing.** Congestion pricing entails charging fees or tolls for road use that vary by level of vehicle demand on the facility. The objective is to bring supply and demand into alignment.

Matching Bottleneck Treatments to Needs

In 2006, as part of the research conducted for National Cooperative Highway Research Program Project 3-83 ("Low-Cost Improvements for Recurring Freeway Bottlenecks"), a series of interviews was conducted with state and local transportation agencies to assess the effectiveness of low-cost improvements used at bottleneck locations within their jurisdictions. The results showed that agencies are using a wide range of strategies to improve bottlenecks, many of them low-cost improvements that can be implemented quickly. The most frequently used operational improvements were ramp metering, auxiliary lanes, and HOV lanes.

Some of the key questions and considerations when selecting improvement alternatives for bottleneck removal include:

- Is there an inside shoulder that would create a usable traffic lane for a short section of freeway?
- If a shoulder is considered for conversion, is there right-of-way (ROW) to allow adding one back for part of the length of the project?
- If there are bridges, are they wide enough to accommodate the extra lane while allowing adequate clearance to barriers (2 feet) and an outside shoulder? If not, are they short enough that a loss of shoulder as a breakdown lane would not be critical (500 feet or less)?
- If changes to an entrance or exit ramp or weaving area are considered, will adjusting the position of ramp gores cause geometric problems which must be resolved?
- Are vertical clearance issues, grade-matching, and sight distance problems created?

- If the bottleneck movement itself cannot be fixed reasonably, can the other traffic which is affected by it be better accommodated?
- Finally, will the improvement invite enough new traffic to cause immediate breakdown again or is this truly the clearing up of a "kink" in the system, without being a capacity addition that will overload some other part of the facility?

Barriers to Implementing LBR Treatments

Because some bottleneck treatments involve innovative solutions that maximize effectiveness with a minimum of new construction, they are occasionally at odds with **highway design standards**. A design exception may be required. For example, the addition of slip ramp to a collector/distributor road or the use of a shoulder as a through lane at selected locations may not strictly follow allowable design standards. Such deviations have the potential to degrade safety if not properly implemented; the elimination of a shoulder may lead to more collisions with roadside features or may impede incident management activities. As it is FHWA's intent to foster creative approaches for low-cost bottleneck improvements, agencies should not see the design standard issue as insurmountable. Rather, they should fully assess the potential safety impacts of strategies and devise ways of addressing them, if necessary. For example, in the case of a shoulder-to-lane conversion, review of crash data, and the specific roadway location (perhaps through a Roadway Safety Audit), it may be determined that a barrier is required to keep vehicles off of the roadside. It may also require a change in incident management policy that would allow emergency vehicles to access incidents from the opposite direction. Finally, agencies should be in contact with the FHWA Division offices throughout the process as design review may be required, depending on circumstances.

The second potential issue relates to **air quality conformity**. Because they are short term in nature, localized bottleneck improvements may emerge as formal projects that have not been previously identified in Statewide Transportation Improvement Programs or MPO-generated Transportation Improvement Programs. Thus, they may not be part of those projects that have been approved to deal with air quality issues in the region or state. Such occurrences must be dealt with on a case-by-case basis by agencies wishing to undertake bottleneck improvements. One point worth noting: if air quality conformity in a location precludes or discourages major capital expansion (e.g., additional lane-miles), the type of improvements in a localized bottleneck program clearly do not fall in this category.

Finally, another potential barrier is that **bottlenecks may not be seen as consistent with Long-Range Design Concepts**. As discussed previously, most bottleneck mitigation strategies such as roadway widening, left-turn lengthening, auxiliary lanes on freeway, or improvement of weave/merge areas may all be seen as distracting resources from larger design solutions, which will be made anyway in a larger longer-term project already in a 20-year plan. Agencies must decide and weigh the benefits themselves whether the cost of doing a smaller bottleneck solutions in the short terms against the cost of waiting for a more complete solution.

Bottleneck Assessment/Analysis Methods

Once bottleneck locations have been identified, the root cause and severity must be determined. Special travel time runs, aerial photography, or video of suspected bottleneck areas can be used to pinpoint sources of operational deficiencies. On freeways equipped with detection technology, dynamic surveillance can be used to identify where and how often bottlenecks occur, and how severe they are. Archived traffic data can be used to measure whether the problem is growing or receding.

Sometimes, the operational cause of a bottleneck is evident, intuitive, or anecdotal. However, when multi-mile corridor congestion is prevalent, microsimulation modeling can assist in identifying, separating, and analyzing bottleneck dynamics within the corridor.

Bottleneck analysis is necessary to study not only the subject location, but also the impacts of potential bottleneck remediation on upstream and downstream conditions. The analysis will justify action to correct bottlenecks, confirm the benefits of bottleneck remediation, or check for hidden bottlenecks along a corridor. When conducting bottleneck analysis, care should be taken to ensure that:

- Improving traffic flow at the bottleneck location does not just transfer the problem downstream. The existing bottleneck may be “metering” flow so that a downstream section currently functions acceptably, but the increased flow will cause it to become a new bottleneck.
- Future traffic projections and planned system improvements are inclusive in the analysis. Safety merits also should be strongly considered.
- “Hidden bottlenecks” are considered. Sometimes, the queue formed by a dominant bottleneck masks other problems upstream of it. Improving the dominant bottleneck may reveal these hidden locations.

It is important to take into account the possibility of "hidden bottlenecks" during the analysis stage.

- Conditions not traditionally considered by models are acccunted for. There are several bottleneck conditions, such as certain types of geometrics and abrupt changes in grade or curvature, that cannot be analyzed by current analysis tools. Engineering judgment will need to be exercised to identify those problems and possible solutions.

For example, **New York DOT's Region 11 LBR Program** is comprised of an interdisciplinary team headed by the Planning Director. The LBR Team compiles various completed studies and will bring other agencies on board as needed in order to develop a current inventory of bottleneck locations citywide; develop screening criteria; select and recommend priority locations for improvements; develop strategies for the selected locations and prepare the implementation plan.

Projects are selected based on the following screening criteria: 1) project cost less than $10M; 2) annual vehicle-hours-delay > 25,000 hours; 3) speeds < 30 mph; 4) requires little or no environmental documentation; and 5) projects NOT already programmed in the TIP.

The LBR Team works with Program Management to address project programming, funding, and schedule issues, and they coordinate with local MPO staff regarding any TIP and/or conformity issues. As the work of the LBR program becomes better defined and develops greater focus, the team will expand to include other external agencies as well.

In **Utah DOT's Choke Point Program**, potential projects are identified through public observations in the regions and input from users and local governments. Potential projects are then selected and prioritized based upon several factors, including average annual daily traffic (AADT), volume to capacity ratio (V/C), constructability, region priority, and accident rates.

Qualifying criteria for potential projects include: 1) small safety, capacity or bottleneck projects that cost less than $10M; 2) can qualify for a categorical exclusion; 3) can be designed and constructed within one year; and 4) require minimum right-of-way needs.

Examples of chokepoint projects include roadway widening, left-turn lengthening, dual left-turns, auxiliary lane on freeway, intersection or signal improvement, passing lanes on rural routes, and improvement of weave/merge areas.

Evaluating Bottleneck Improvement Projects

After implementation, it is often beneficial to conduct an "after" evaluation to gauge the effectiveness of the bottleneck removal project. A conservative approach to evaluating treatment effectiveness is to evaluate operational and safety benefits achieved. Common evaluation methods include microsimulation modeling, benefit/cost analysis, and crash data analysis using data collected before and after project implementation. In addition, the following performance measures are often used to assess the effectiveness of bottleneck improvement strategies: average speed (travel time), lane density, queue lengths, queue discharge rates, vehicle miles of travel (VMT), and vehicle hours of travel (delay). Additional insight could be obtained from before and after opinion surveys of area drivers. These types of evaluations are often not done, yet are important to quantify the benefits achieved through bottleneck mitigation.

Bottleneck Performance Measures in Bottleneck Assessment and Evaluation

This section describes bottleneck performance measures, data and methods needed to develop the metrics, presentation of performance measure results, and matching bottleneck treatments to needs.

Metrics

There are a number of metrics, which could be used to identify and track the results of bottleneck mitigation strategies. Below is a list developed by the Mn/DOT.

- Percentage of miles congested (i.e., number of directional miles with speeds < 45 mph during peak periods);
- Total daily delay (volume x time difference between actual and posted speed);
- Reliability (TTI – Congestion Index/Travel Time Buffer Index);
- Miles of FIRST/incident response coverage;
- Number of hours where volume > capacity;
- Frequency of signal retiming;
- Percent of arterials with coordinated signals;
- "Before/After" benefit/cost ratio of corridor improvements;
- "Before/After" benefit/cost ratio of signal retiming;
- Customer satisfaction survey of peak-hour travel (omnibus transportation survey);

- Percent of MUFS instrumented;
- Planned lane closures System;
- Unplanned lane closures System;
- Average clearance time for snow and ice removal;
- Average clearance time for freeway incidents; and
- Throughput (i.e., number of vehicles through a specific corridor or across a screenline over a specified time period).

Data and Methods Needed to Develop Metrics

Data collection to support bottleneck analysis should be sufficient to capture the duration and extent of congestion. Typically, 15-minute traffic volume counts for all ramps and main lanes for a four-hour peak period are adequate. Other data can be collected through travel time runs, video, or origin-destination studies. Many regional TMCs are archiving the data from their systems and can be a very helpful resource.

Presentation of Performance Measures

Performance measures can be presented in a number of formats. The one recommended for this effort include routine reports on an annual or semiannual basis. If more frequent reporting is required, many emerging on-line reporting tools are becoming available that provide robust dashboard style results to easily convey to leadership and the public the status of a bottleneck program.

2.4. Bottleneck Self-Assessment

Table 2.1 describes procedures for agencies to conduct an LBR Self-Assessment to establish where they currently stand with regard to implementing a bottleneck program, perhaps to the level of identifying at which "stage" it is.

2.5. Templates for Getting Started

This section introduces three templates that transportation agencies can consider in developing their own LBR Program. Each template includes an overall description of the approach, a bullet list of the key elements of the approach, followed by a step-by-step guide to implementing the template.

Table 2.1. Bottleneck Programming Maturity Model Matrix

Level	0 – Ad Hoc	1 – Aware	2 – Planning	3 – Defined	4 – Managed	5 – Integrated	6 – Continuous Improvement
Institutional	No structured LBR Program in place.	Some level of bottleneck recognition.	Agency is discussing needs/plans for a structured LBR program.	Development of a formal LBR program is underway.		A formal LBR program is named and established, either as a separate program or part of another program.	
Organizational	No LBR Program Champion or Team in place.	Agency is aware of the need for a LBR Program champion and Team to support LBR activities.	Some level of LBR program assessment and formulation of roles for LBR Team is underway in one or more offices of agency.			Specific staff have the responsibility for oversee a formal LBR program.	
Planning/ Programming	LBR strategies are not included in agency planning/programming processes.	Agency has developed procedures for considering LBR strategies in the planning and programming processes.		LBR strategies becomes a component of the CMP.	LBR strategies integrated with short- and long-term planning and programming.	LBR strategies appear in planning and programming documents: MPO TIP, LRTP, annual programmed projects list.	
Performance Measures	No defined practices and techniques for bottlenecks.	Agency has produced a set of procedures for measuring performance of bottlenecks (evaluation procedures).		Selected LBR projects are evaluated.	LBR projects are routinely evaluated.	LBR projects are highlighted in agencywide reports of congestion and safety.	

Level	0 – Ad Hoc	1 – Aware	2 – Planning	3 – Defined	4 – Managed	5 – Integrated	6 – Continuous Improvement
Assessment/ Analysis Tools	No defined tools to support bottleneck assessment/ analysis process.	Agency has defined procedures for identifying and assessing LBR strategies.				A combination of analytical methods are used at different stages of LBR assessment and evaluations.	

Finally, each template concludes with a real-world example of the approach discussed.

The following templates are included in Appendix A.

- Template #1: Public Outreach Identification Process;
- Template #2: Leverage Existing Non-Bottleneck-Related Processes; and
- Template #3: Internal Identification Processes.

2.6. Sustaining a Successful Program

Once a LBR program has been established, its ongoing success will be based largely on two factors, funding and institutional integration. If funding for LBR programs is either dedicated or at least available through traditional sources, the chances that LBR issues will be considered in the future is very high. If, however, an agency goes through the initial effort of setting up a LBR program and is at the end of the process unable to secure dedicated funding, or at least access to traditional sources, the program will not remain viable. It is, therefore, critical that agencies consider funding options early on in the process. That is not to say LBR effort should not be initiated without funding secured; just the opposite. Many times the effort of setting up the LBR program can build enough momentum within an organization to secure funding or at least ensure access to compete with for funding.

Likewise, sustaining an LBR program will be much easier if the program is integrated into an organization. If identifying and evaluating LBR projects can become part of the core mission of the agency, the LBR program will continue even after the initial champions have left the organization or been reassigned.

2.7. Signs of a Latent Program

It may be prudent here to take a moment and discuss "when is a program *not* a program," that is to say, when a program has de-evolved to not-producing effective mitigation projects.

Certainly, there are some administrative actions that would annul any active program. Foremost among these would be budget cuts, new administration initiatives, and emergencies. But in context of recognizing a dormant localized congestion mitigation effort, here are some vital signs to check.

- **Projects are not being produced under "low-cost, short timeframe" measures.** The LBR program specifically does not define these terms in the context of the program; however, if projects are not being completed within "annualized" budgets, or are not being delivered in relatively short timeframes (e.g., a "construction season" as opposed to multiple out-years) then chances are these projects are outside the de facto definition of a "localized" congestion mitigation project.
- **Candidate project lists are drying up.** If a pre-standing list of candidate mitigation locations either does not exist or has not been updated for some time, then it may be a sign that program momentum has lapsed.
- **Is the work effort generally equitable to the project size?** Generally speaking, low-cost, short-term projects should have a much smaller planning and execution effort than large projects. By their very nature, these projects should be quick turn-around, low-cost, and high-yield benefits. However, if the project is bogged down with overly lengthy studies ("paralysis by analysis"), add-ons, or cost escalations, then it may be a sign that said project is beyond the scope of "localized."

All of the above signs of a latent program are further discussed in FHWA's "Traffic Analysis Toolbox X: Localized Bottleneck Congestion Analysis" (FHWA-HOP-09-042) which focuses on what level of analysis is appropriate and representative for the subject projects and programs. In summary, that document concludes that the level of analysis (of any problem) should roughly correlate to the size of the problem. **The main questions that this guidance helps an agency frame are: 1) do we have a satisfactory agency methodology to *specifically* address localized congestion problems; 2) within that program, do we have satisfactory justifications for project candidacy, selection, and solutions for said problems; and 3) are we executing these projects in a timely fashion and within budgets that are representative of the context of a "Localized Bottleneck Reduction" program?**

3.0. ADDITIONAL RESOURCES

3.1. Case Studies of Agency Programs

The following provide comparisons of how different state agencies have incorporated low-cost bottleneck projects into their planning and programming processes.

Caltrans

The California Department of Transportation (Caltrans) does not have a formal bottleneck planning process; rather, bottleneck issues are addressed at the district level as part of their Corridor System Management Plans (CSMP), which are developed for some of California's most congested transportation corridors. System monitoring and evaluation is seen as the foundation for the entire process because it cannot only identify congestion problems, but also be used to evaluate and prioritize competing investments.

The CSMP includes the identification of bottlenecks and potential short-term fixes as part of an overall and long-term strategy for making corridor improvements.

This may take the form of an "LBR audit," which is a review of traditional large-scale corridor studies to identify opportunities for using LBR improvements as part of the package of improvements. The LBR audit concept is similar to that of Road Safety Audits.

Caltrans does not have a direct funding for bottlenecks, although bottleneck projects are routinely programmed through the CSMP process.

Florida DOT

In Florida, there is not a "bottleneck" planning process, per se; rather, bottleneck-related issues are addressed as part of the Florida Department of Transportation's (FDOT) standard planning process.

The planning process, which is managed by the FDOT Systems Planning Office, begins with needs identification conducted at the district level, then projects are developed and proposed for the Cost Feasible Plan.

The Cost Feasible Plan is adopted and projects are ranked for inclusion into the 5- or 10-year programs. Traffic data and the statewide model are used to identify deficiencies, but it is the responsibility of the districts to identify and resolve hot spots.

Indiana DOT

Indiana DOT is in the beginning stages of developing a bottleneck program, as they are transitioning to a new asset management-based funding system with multiple funding teams. A program champion has been identified. Funding for bottlenecks will fall under the newly formed Mobility Asset Funding Team. The program will initially focus on Interstates, although INDOT is still considering how to best identify the spot location bottlenecks. It is expected that bottleneck countermeasures will have a very good benefit/cost ratio and compete well in the funding decision matrix that is a part of the new funding system.

Maryland State Highway Administration

The Maryland State Highway Administration (SHA) has a dedicated program of about $5M per year for the identification and implementation of low-cost traffic congestion improvements at intersections. The program's genesis tracks to when SHA asked, "What can be done if and when a megaproject's 'no-build' alternative is chosen?" The program has been well received by the public and local governments. Projects typically include low-cost projects that can be implemented quickly, such as signal timing upgrades and adding turn lanes and through lanes at intersections. The Maryland SHA also has had considerable success with projects to improve freeway ramps and merge areas that have reduced congestion bottlenecks at a low cost.

Michigan DOT

Michigan DOT currently is in the process of developing a structured Localized Bottleneck Reduction Program. The effort began several years ago with structured changes at MDOT, during which MDOT officially reorganized their Maintenance and Traffic and Safety Divisions to create a Division of Operations. The next step was the formation of a new section titled Systems Operations and Management (SOM). One of their early charges was to develop an approach to identify and eliminate bottlenecks throughout the State. Several years previous to this reorganization, MDOT developed and utilized a "Choke Point" Program, and their current efforts are patterned after that effort.

One of the first official action steps that the SOM Section pursued was to solicit potential bottleneck locations and problem descriptions from each of their seven region offices. More than 200 locations were identified, with about one-third being freeway interchanges. Based on further review by the SOM

Section, the total number of potential locations was reduced to approximately 125 locations, which they believed: 1) met their definition of a "bottleneck" location; and 2) had a potential cost-effective solution that could address the problem. One of the primary goals of this highly focused initial effort is to develop a documented and sustainable approach that can demonstrate excellent benefit-to-cost ratios, as well as justification for allocation and expenditure of funds on the statewide LBR Program. The underlying goal was to obtain leadership support and a dedicated funding template specifically for bottleneck reduction projects, which has now been achieved.

Many challenges exist as the Program and structure move forward. One primary challenge is the need to complete a detailed analysis necessary for a large number of potentially competing projects, as well as a freeway analysis of these projects. MDOT staff resources are limited and MDOT is reviewing the potential use of consultants and/or universities for project analysis. Another issue is how to justify and evaluate the impacts of the suggested changes as well as the existing problem. The intent is to create a level playing field for application of LBR funding by each of the seven regions. The third major challenge is the availability of funding. Michigan is going through an extremely dynamic period with the overhaul of the automobile industry, and their funding has been reduced. These issues are all being discussed and debated as MDOT moves forward to establish and document a formal, fully funded LBR Program.

Minnesota DOT

Minnesota DOT was originally driven to explore low-cost congestion relief projects because of budgetary restrictions, but soon realized that these projects could be implemented very quickly and, as a bonus, were highly visible and popular with the public. In much less than one year, Mn/DOT developed a highly accelerated process for bottleneck identification and prioritization, which led to many effective projects in the following two years. Mn/DOT also found that because of lower costs, it could identify multiple locations throughout the region and "spread around" bottleneck reduction projects in a fair and equitable manner. This process consisted of completing a study, which included a five-step process to narrow potential projects into a recommendation list to the state legislature. Evaluation of completed projects produced high benefit/cost ratios, usually greater than 8:1. *Note: Circa 2009, this one-time activity was replaced by an ongoing CMS process known as the Congestion Management Planning Process, which has been formally adopted as part of the 3C planning process.*

Minnesota's Process to Identify and Prioritize Bottleneck Improvements

Step 1: Project Identification
Potential congestion management projects were identified from existing sources:

- Low-cost capacity improvements (e.g., auxiliary lanes);
- Restriping lane configuration; and
- Traffic control device improvements (e.g., ramp meters and signal timing).

Step 2: Quantitative Screening

- Project cost < $15 million
- Not in three-year TIP
- Annual hours of delay > 25,000
- Minimum of two hours of congestion

Step 3: Qualitative Screening

- Design readiness
- Cost range
- Congestion benefit
- Construction traffic management
- Future demand changes
- No adverse downstream effects

Step 4: Expert Workshop
Projects were prioritized by an expert group during a half-day workshop.

Step 5: Project Planning
The following were prepared for each project:

- Geometric sketches;
- Project scope;
- Congestion impacts;
- Safety impacts; and
- Benefit-to-cost ratio.

New York DOT

New York DOT's Regional 11 Planning Office (New York City) established their Localized Bottleneck Reduction (LBR) program with a mission to mitigate recurring congestion at selected chokepoints and the operational influences that cause them on the region's highway corridor mainlines, service roads, ramps, and immediate adjacent local streets and intersections. The operational influences include highway junctions and decision points, lane drops, weaves, merges, ramps, signals, intersections, width-restricted underpasses, and other factors that can adversely impact traffic flow at moderate to high volumes. The LBR program will investigate the opportunities as well as develop measures for the application of operational and low-cost infrastructure improvements to address chokepoints and to identify cost-effective improvements either as standalone initiatives or as part of existing Capital Projects. The goal of the LBR will be to move recommendations into the capital program.

Project Team. The LBR Program is administered through an interdisciplinary team headed by the Planning Director, with members designated by the functional unit managers. The approach is modeled after the Preliminary Feasibility Investigation (PFI) concept currently used by the Planning Office, which utilizes a multifunctional project scoping team as the "vehicle" for building consensus on the nature of a transportation problem, the key issues, and best solutions. This approach has been used by the Planning and Development unit to focus on and analyze the feasibility of specific projects, and it creates a forum for early consensus and buy-in of preferred alternatives and course of actions by the different functional groups.

The LBR Team will evaluate various completed inventories/studies and will bring other agencies on board as needed in order to develop a current inventory of bottleneck locations citywide; develop screening criteria; select and recommend priority locations for improvements; develop strategies for the selected locations; and prepare the implementation plan. The team will work through Program Management to address programming, funding, and schedule issues, and they will coordinate with the local MPO regarding any TIP and/or conformity issues. At the present time, the Localized Bottleneck Reduction program will be confined only to Region 11 functional groups plus FHWA. As the work of the LBR program becomes better defined and develops greater focus, the LBR PFI Team will be expanded to include external agencies, such as the NYCDOT, NYCT, MTA, PANYNJ, etc.

Program Benefits. The benefits of the LBR program include:

- Proactively engaging in reducing recurring delays and improving quality of life for affected communities;
- Reduce the rate of RCNs by elected officials and citizens to the department;

- Low-cost, effective solutions which will not require significant environmental actions and are relatively low-cost capital projects or maintenance work orders;
- Short implementation schedule; and
- Provide opportunity for incorporating recommended mitigations into ongoing projects, etc.

Screening Criteria and Schedule. The screening criteria for project selection is as follows: 1) project cost < $10M; 2) annual vehicle-hours-delay > 25,000 hours; 3) speeds < 30 mph; 4) requires little or no environmental documentation; and 5) projects NOT already programmed in the TIP. Once the LBR PFI Team is constituted and established, it is expected that within the next 9 to 12 months, the identification and assessment, including corresponding mitigation strategies, of up to four bottleneck locations, will be available for further LBR team review and comment.

Ohio DOT

In 2008, Ohio DOT initiated a new Systematic Signal Timing Program that provides district offices and local governments with technical assistance (task order consultants) to analyze, coordinate, and upgrade the timing and phasing of signal systems in high-crash areas. Funding can also be used to upgrade signal equipment as needed. Funding is targeted to those areas listed as a top crash priority by ODOT, MPOs, or local governments. Project sponsors must verify that the crash patterns are relevant to signal timing concerns. Requests are funneled through the appropriate ODOT district office, which typically must concur with the request for assistance.

In Ohio, bottlenecks are part and parcel of the overarching Ohio Department of Transportation (ODOT) Highway Safety Program (HSP), which ranks all candidate projects and drives the statewide highway project selection and scheduling process. Beginning in 2002, ODOT developed a "congestion mapping" division that uses V/C ratios developed from traffic data recorders and roadway inventory. About the same time, ODOT administration pushed for an annual process of overlaying congestion-index and safety-index "hot spots." As a result, congestion hot spots now have a "voice" in the process regardless of crash indices, and congestion-related problems now compete for attention in the HSP listing. Specifically, highway sections with V/C ratios greater than 1.0 are considered "congested" and are added to the listing. Sections with V/C between 0.9 and 1.0, but outside the cities of Columbus, Cincinnati, and Cleveland, are also added. After ODOT headquarters completes their statewide effort of congestion mapping and safety indexing, the respective District engineers are responsible for developing countermeasures for their top-listed candidate projects. District Safety Review teams sort projects into three scales – low (less than $100K and quickly implementable), medium ($100K to $5M and one to two years), and high (greater than $5M and necessitating more than years to implement) – and then compete with other projects having the same scale but in other districts.

Utah DOT

Funded by the Utah Legislature, Utah DOT initiated a Choke Point Program in 2006 to address safety, congestion, and bottleneck areas. It was successful, so the Legislature put forth more funding for additional chokepoint projects. The UDOT has used the choke point program to identify potential projects that could improve operations and safety with relatively low cost and in a short time period. Potential projects are identified through public observations in the regions and input from users and local governments. Qualifying criteria for potential projects include small safety, capacity, or bottleneck projects that: 1) cost less than $10M; 2) qualify for a categorical exclusion; 3) can be designed and constructed within one year; and 4) require minimum right-of-way needs. Examples of choke point projects include roadway widening, left-turn lengthening, dual left turns, auxiliary lane on freeway, intersection or signal improvement, passing lanes on rural routes, and improvement of weave/merge areas. Projects are selected and prioritized in each region based upon several factors, including Average Annual Daily Traffic (AADT), volume-to-capacity ratio (V/C), constructability, region priority, and accident rates.

Virginia DOT

Virginia DOT implemented the Strategically Targeted Affordable Roadway Solutions (STARS) Program. This program is a safety and congestion program that partners state, planning district and local transportation planners, traffic engineers, safety engineers, and operations staff to identify "hot spots" along roadways where safety and congestion problems overlap and are suitable for short-term operational improvements. The major goals of the STARS Program are to identify roadway improvements that:

- Are relatively low in cost with no more than $2M for a primary project and $5M for an Interstate project;
- Address existing mobility and safety problem areas;

- Require minimal preliminary engineering and right-of-way; and
- Can be implemented quickly, with a goal of 24 months or less.

Current funding for the program is through the Highway Safety Improvement Program (HSIP), although VDOT is looking for other funding mechanisms.

STARS Process
Step 1. Study Area Selection Identify critical safety hot spots, and overlay congestion hot spots Step 2. Conduct Detailed Study Process Conduct objective and quantifiable study process Step 3. Prioritize Recommendations Prioritize recommendations based on benefit/cost analysis results Step 4. Programming and Implementation

Washington State DOT (WSDOT) has no direct funding or a separate program just for bottlenecks, but formally recognizes "bottlenecks and chokepoints" in their project planning and development process and devotes a portion of the Washington Transportation Plan (WTP) to them. Financing large congestion relief projects has become a serious problem due to reduced tax revenues, so WSDOT has organized the *Moving Washington* initiative. As stated in their most recent Congestion Report:[2]

> ... WSDOT applies three balanced strategies to fight congestion – operate efficiently, manage demand, and add capacity strategically. By strategically adding capacity, WSDOT targets bottlenecks and chokepoints in the transportation system. However, because of limited resources, WSDOT understands that adding capacity cannot be the only solution for solving the congestion problem. That is why WSDOT uses operational strategies to maximize the efficiency of the existing transportation system (operate efficiently). WSDOT manages demand by providing alternatives to drive-alone commutes between and within modes of travel and encouraging the traveling public to use them. Performance results show that Moving Washington strategies and projects are making a difference around the State to relieve congestion.

Strategic, low-cost capacity improvements currently are a major emphasis for WSDOT. In the safety area, WSDOT learned that by targeting "black spots" rather than corridor- or systemwide safety improvements, much greater impact for investments could be achieved. This same philosophy is being carried over to bottlenecks.

The *Moving Washington* initiative is part of the larger planning and programming activity of WSDOT. The Highway System Plan[3] is the result of this process, which is performance-driven, as described below.

The process starts with definitions. Bottlenecks and chokepoints are defined differently from a functional viewpoint, but they both relate to the physical characteristics of the facility that impede traffic. Bottlenecks are places where the physical attributes of a roadway change in a manner that impacts the flow of traffic. Typical bottlenecks are locations where the number of lanes decreases; the roadway physically narrows either in shoulder width or lane width and narrow bridges. WSDOT defines chokepoints as places where congestion occurs because of traffic interference and/or the roadway configuration. Both bottleneck and chokepoint locations can be identified by WSDOT or other agency staff just by observation.

The next step is that the observed congestion must be supported with traffic data and analysis models. If congestion is a problem today or anticipated within the next 20 years, it also must satisfy one of the following applicable criteria:

- The congestion problem impacts the flow of mainline through-traffic. Mainline traffic flow is considered to be impacted when through-vehicle peak-hour speeds are equal to or less than 70 percent of the posted speed.
- Traffic flow criteria for ramps will also be considered to determine if the congestion is caused by on/off-ramp traffic.

Once problem locations are identified, potential improvements are identified by central WSDOT staff, regional WSDOT staff, and MPO staff. The potential improvements follow a tiered process:

- First, low-cost improvements are considered, including LBR-type projects and operations.
- Second, larger-scale projects with a longer duration of benefits, short of major reconstruction, are considered.
- Finally, the "ultimate" long-term fix is identified, which usually involves major reconstruction.

Problem locations can have one, two, or three of these tiers identified for it. Solutions are then tested using analytic models developed by WSDOT for benefit/cost ratios and performance analysis. From this, the best combination of improvements is identified statewide.

The entire planning and project development process is performance-based. This fits into WSDOT's longstanding emphasis on using performance measures to guide investment as well as the Governor's Government Management, Accountability and Performance (GMAP) program, which spans all state agencies. Every month, the Governor and her staff meet with the heads of state agencies and departments to evaluate the performance results that these organizations currently are delivering. These open, candid meetings provide insight into what is and what is not working in state agencies, and what is needed to improve performance and achieve on deliverables. These meetings (held in Olympia twice a month in the Legislative Building) are open public forums.[4]

Metroplan, Little Rock, Arkansas

Metroplan, the MPO for the Little Rock region, has implemented a program dubbed "Operation Bottleneck" aimed at identifying current congested locations that are amenable to relatively quick and inexpensive treatments. Major congestion problems – arterial corridors and freeway sections/interchanges with major capacity deficiencies – are well known throughout the area. Further, future (major) problems have been identified with the modeling done for the long-range transportation plan. However, funding for the major improvements necessary at these locations must come from either:

- State DOT, Arkansas State Highway and Transportation Department (they would be managed as state projects; and
- Local governments saving up several years of state and Federal allocations for a single project.

Metroplan wanted a way to serve their constituents better than constructing a scarce few megaprojects. Further, the region is almost in nonattainment for the eight-hour ozone standard. The text from their press release on the program sums up their intent very well:

> "We're aware of the major congestion issues in our area and have identified those in our long-range plans, but we know there are dozens,

maybe hundreds of neighborhood problems throughout the region that could be fixed with something as simple as a roundabout or coordinating traffic signals to improve flow," McKenzie says. "Localized problems like these can be harder to identify and are sometimes overlooked, even though they can be just as frustrating to drivers. Often they can be addressed much more quickly than larger projects. Those are the types of areas we are hoping to identify through Operation Bottleneck."

Operation Bottleneck is largely based on the establishment of a Regional Mobility Authority (RMA). In Arkansas, an RMA is a coordinating body with no taxing powers – member counties would have to raise the taxes necessary to fund projects; multiple counties would be involved. Most likely, the RMA will be based on a temporary increase in local sales taxes county-by-county; they feel it is important to sunset the tax so it is more palatable to the public and elected officials. Metroplan hopes to leverage state and Federal funds against their self-generated revenue to fund the projects.

Also key to the strategy is a specific list of projects to be funded by the tax increase, and most of the Operation Bottleneck effort has gone into project identification, as discussed below.

Project identification is being driven almost exclusively by public input via local meetings and an Internet survey. Metroplan also hired a marketing firm to promote the program through local media. Both congestion safety problem areas are being solicited, along with other modal deficiencies (transit, special transportation).

A huge range of responses has been received, from megaprojects to minor problems on local roads. For congestion problems, signals and interchanges are dominating the responses. Safety problems identified by the public tend to be more general than site-specific. (This is understandable since congestion is experienced routinely but crashes are rare events for individuals.)

Metroplan staff will assemble the projects and will develop a list of projects to iterate with the public. Staff will also make revenue projections under different sales tax rates. No formal benefits assessment is planned – as with project identification Metroplan emphasized that public input is the driver for Operation Bottleneck, not technical processes (which they use for all other transportation planning activities).

The staff will compare public-identified projects with those in the TIP and LRTP as well as against congested sections identified in their Congestion Management System in developing a prioritized list. Metroplan staff offered two types of improvements that are likely to dominate the project list:

- Low-cost arterial improvements – improved timing, intersection approach geometric improvements, and access management; and
- Roundabouts at uncontrolled, stop sign-controlled, or low volume signal locations.

Initial results are planned to be presented to public officials in October 2008.

Metroplan would like to make this an ongoing process, especially since the public support for the program has been very high. How to structure the funding for an ongoing program will be tricky, however.

A. TEMPLATES FOR IMPLEMENTATION

A.1. Template #1. Public Outreach Identification Process

Overview

This approach to starting a LBR program utilizes a public outreach campaign to identify bottleneck locations. After all, who knows more about traveling local roads than the people who use them every day? Major congestion issues are usually thoroughly identified through the traditional regional long-range planning approach, but there are hundreds of more targeted neighborhood-based problems throughout any region that are sometimes overlooked, even though they can be just as frustrating to drivers. Often these bottlenecks can be addressed much more quickly than larger projects.

Key Elements

- Good method to quickly identify bottlenecks at a modest cost.
- Requires coordination between different organizational elements within an agency.
- Builds public support which could be leveraged for targeted revenue initiatives to fix the worst bottlenecks.
- Response rates from the public can be significant and feedback of results is critical to success.

Timeframe	Cost	Institutional Hurdles	Analytical Complexity
Quick Turnaround	Moderate	Moderate	Low

Approach

Step TI. Establish the Types of Bottlenecks to Solicit

Agencies should first decide how broadly or narrowly they want to explore bottlenecks in their region. For example, a region could choose to only look for suggestions for pedestrian-related issues. Likewise, the focus could be on signalization issues on corridors or even centered on modal concerns.

Step 2. Determine Outreach Mechanisms

Once the scope of the outreach is clarified, an approach to solicit the public's input should be developed next. Leveraging the public input is an appealing approach to identifying bottlenecks since almost all transportation agencies have experience in communicating with the public. Outreach mechanism to consider include public meetings, telephone surveys, mail-in questionnaires, web-based surveys, and even newer social media surveys.

Whichever approach is selected, a targeted questionnaire should then be developed. The length, detail, and composition will be driven by the delivery mechanism. Care should be taken if multiple channels are utilized such as public meetings and telephone surveys. In these cases, one survey approach or questionnaire will probably not be appropriate and multiple methods to solicit responses to the public will be warranted.

Agency Public Information Offices (PIO) should be engaged in this process early; since they can provide a wealth of experience and expertise in reaching the public. The PIO can also help secure local press coverage of the agency's bottleneck efforts. This exposure can increase participation in public meetings as well as any survey mechanisms utilized.

Step 3. Develop Prioritization and Sorting Methodology

Once bottleneck locations are collected, a mechanism must be developed to rank and prioritize the responses. Transportation agencies routinely prioritize and rank projects. However, for this effort, it is not recommended that any traditional extensive modeling or benefit analyses be conducted. These traditional planning processes, although well established, are too cumbersome for this effort since in this case input from the public is the driver not a technical processes.

Instead, priorities should be established by agencies using sketch level and simplified ranking mechanisms. Examples in include comparing the public identified projects with those in the TIP and LRTP as well as against congested sections identified in their Congestion Management System.

Priorities could also be assigned by simply assessing the frequency of the number of occurrence of either specific bottlenecks or bottleneck subregions thereby generating regional hot spots.

Step 4. Create Feedback Mechanism

Sharing the results of the bottleneck outreach and prioritization is vital to the success of this approach. Working with the PIO, agencies should ensure the results are released. Examples include holding another round of public meetings, additional press conference or releases, or a coordinated public relations campaign to promote the findings.

Example: Operation Bottleneck, Metroplan, Little Rock, Arkansas

Metroplan, the MPO for the Little Rock region, implemented a program dubbed "Operation Bottleneck" aimed at identifying current congested locations that are amenable to relatively quick and inexpensive treatments. Major congestion problems – arterial corridors and freeway sections /interchanges with major capacity deficiencies – are well known throughout the area. Further, future (major) problems have been identified with the modeling done for the long-range transportation plan. However, funding for the major improvements necessary at these locations must come from either:

- State DOT, Arkansas State Highway and Transportation Department (they would be managed as state projects); and
- Local governments saving up several years of state and Federal allocations for a single project.

Metroplan wanted a way to serve their constituents better than constructing a scarce few megaprojects. Further, the region is almost in nonattainment for the eight-hour ozone standard. Operation Bottleneck is largely based on the establishment of a Regional Mobility Authority (RMA). In Arkansas, an RMA is a coordinating body with no taxing powers – member counties would have to raise the taxes necessary to fund projects; multiple counties would be involved. Most likely, the RMA will be based on a temporary increase in local sales taxes county-by-county; they feel it is important to sunset the tax so it is more palatable to the public and elected officials. Metroplan leverages state and Federal funds against their self-generated revenue to fund the projects. Also key to the strategy is a specific list of projects to be funded by the tax increase, and most of the Operation Bottleneck effort went into project identification, as discussed below.

Project identification is being driven almost exclusively by public input via local meetings and an Internet survey. Metroplan also hired a marketing firm to promote the program through local media. Both congestion safety problem areas are being solicited, along with other modal deficiencies (transit, special transportation). A huge range of responses have been received, from megaprojects to minor problems on local roads. For congestion problems, signals and interchanges are dominating the responses. Safety problems identified by the public tend to be more general than site-specific.

Metroplan staff assembled the projects and developed a list of projects to iterate with the public. Staff also made revenue projections under different sales tax rates. No formal benefits assessment was conducted – as with project identification Metroplan emphasized that public input is the driver for Operation Bottleneck, not technical processes (which they use for all other transportation planning activities). The staff compared public-identified projects with those in the TIP and LRTP as well as against congested sections identified in their Congestion Management System in developing a prioritized list. Metroplan staff said two types of improvements that dominated the project list:

1) Low-cost arterial improvements. Improved timing, intersection approach geometric improvements, and access management.
2) Roundabouts at uncontrolled, stop sign-controlled, or low-volume signal locations.

A.2. Template #2: Leverage Existing Non-Bottleneck-Related Processes

Overview

This approach to starting a LBR program promotes integrating a bottleneck program with congestion or safety programs already active in a transportation agency. Since starting a LBR program can sometimes face institutional, bureaucratic, as well as funding challenges, one successful method of addressing bottlenecks is to ensure LBR issues are integrated into already existing agency project identification and funding procedures. Established congestion or safety programs routinely survey for issues and hot spots and provide a great opportunity to be integrated with LBR issues. The advantage of this approach is that LBR-related projects will have access to an established funding mechanism.

Key Elements

- Provides LBR-related projects access to established funding mechanism;
- Bottleneck identification is linked to traditional processes which are potentially slower than other LBR approaches discussed in this report;
- Requires internal institutional reconfiguration of traditional processes which could take time as well as political momentum; and
- Care must be taken in ensuring LBR projects are given opportunity to compete fairly against other more traditional projects.

Timeframe	Cost	Institutional Hurdles	Analytical Complexity
Quick Turnaround	Low	Low	Low

Approach

Step TI. Identify Existing Program

Transportation agencies looking to integrate LBR projects into existing procedures should first do an assessment of which of their current programs are applicable. Options could include integrating LBR concepts into an agency's traditional corridor improvement process. For example, the identification of bottlenecks and potential short-term fixes could be included as part of an overall and long-term strategy for making corridor improvements. Or a review of existing traditional large-scale corridor studies could identify opportunities for using LBR improvements as part of the package of improvements similar to that of a road safety audit. Agencies could also work to integrate LBR activities and initiatives into their existing Congestion Management Programs. Finally, LBR projects could be linked with safety programs and bottleneck issues and criteria could be added to the process of identifying safety projects.

Step 2. Integrate Bottleneck-Related Metrics into Project Evaluation Procedures

Ensuring bottleneck-related metrics are included into the project evaluation process is a critical step in this approach. The metrics to include vary greatly depending on the type of traditional program. However, there are some core metrics which should be considered. Examples include simple V/C ratios, hours of delay, total daily delay, reliability, or something as simple as frequency of signal retiming. Since this approach is promoting the integration of LBR issues into existing processes, it should be noted that only a few bottleneck-related metrics will be eventually included in the final process.

Example

When the Ohio Department of Transportation (ODOT) revamped their Highway Safety Plan (HSP) efforts a number of years ago, one of the controlling elements identified was the level of congestion. A number of analyses showed that the highway safety and levels of congestion were intertwined, so an element was included in the HSP to specifically address congested locations regardless of the crash levels. ODOT's HSP for the past six years has addressed congested locations in addition to freeway and nonfreeway high-crash locations. The HSP is administered by the Office of Systems Planning and Program management in the Division of Planning and the Districts through the District Safety Coordinators. Each year about $65M in funding is awarded for projects in the various categories that make up the program. Additionally, the districts utilize a portion of their budgets for low-cost improvements. The program has four elements in addition to the identification and study of high-crash locations that are described below.

1) **Hot Spot Locations.** Defined as any two-mile segment of freeway with more than 250 crashes or a nonfreeway location with more than 250 crashes over three years.
2) **Rear-End Hot Spot Locations Map.** Crash threshold set at 150 rear-end crashes for the section (both freeways and nonfreeways).
3) **Congestion.** Identified by calculating a roadway's volume to capacity ratio (V/C). Sections with V/C ratios greater than 1.0 are considered congested and added to the annual work plan. Sections with V/C ratios between 0.9 and 1.0 are added if they are outside of Columbus, Cincinnati, and Cleveland.
4) **Corridor Safety Program.** Corridors with the highest density of fatal crashes are studied and addressed using a crossjurisdictional approach that combines engineering, enforcement, and educational resources.

The District Safety Review teams are required to study the locations identified in these components, as well as high-crash locations. The congested locations often show up on multiple lists, including the Hot Spot or High-Crash Location lists. When this is the case, the higher order study takes place and the location is listed as the higher order need. In general, this order is High-Crash Location, Hot Spot Location, then Congestion Location. This makes it difficult to identify all congestion relief efforts since a significant number of locations are listed on the higher order lists. The statistics presented here are based on locations flagged in ODOT's Safety and Congestion Work Plan as congested locations, so they should not be considered as all inclusive of ODOT's efforts in congestion relief.

The list of congested locations is determined using ODOT's Road Inventory databases and applying highway capacity calculations to determine V/C ratios. The methodology includes analysis of signalized intersections and the development of artificial signalized intersections in urban areas that do not have accurate traffic signal inventories. Attachment 1 is the current ODOT list of congested locations that was generated in July of 2007. Study outcomes and fixes are tracked in a database. This database includes all locations since the effort started approximately six years ago.

Each location is studied using an abbreviated format that includes verification of the actual congestion issue by the District Safety Review Team. This is necessary since the data used to determine that statewide list is system level and may not always reflect actual conditions within a given section of roadway. ODOT has been collecting cost data and hopes to eventually be able to evaluate the cost-effectiveness of the various countermeasures being utilized at congestion locations.

Solutions are grouped according to length of effectiveness and cost. Each Congestion Fix receives a Cost/Time code that is determined by rating Cost as Low, Medium, or High and Time as Short Term, Mid Term, or Long Term. The solutions range from simple maintenance activities to the programming of major projects. A number of Short-Term/Low-Cost recommendations have included meeting with the local law enforcement agencies, review of signs and pavement markings for upgrade, and use of time lapse video to analyze actual travel patterns.

Additional low-cost improvements have included signal timing revisions; minor striping and signing changes (e.g., changing lane assignments on intersection approaches); and other measures that can be accomplished by state or local staff. Medium-cost improvements have included signing, striping, and RPM upgrades/revisions; minor widening (e.g., turning lanes; and traffic signal system upgrades). High-cost improvements are either new projects with the primary purpose of addressing the congestion or are major components of larger projects that are often major system rehabilitation efforts.

In summary, Ohio has been studying congestion sections and locations as part of their Highway Safety Program for the past six years and has been applying solutions to these that range from increased law enforcement and roadside assistance patrols to major reconstruction projects. They are collecting data and intend to use it to develop benefit/cost factors for future use and to evaluate the effectiveness of the solutions implemented. The location studies are prepared and reviewed by the District Safety Review teams that are multidiscipline teams well acquainted with using a wide variety of methods to implement solutions. They also can quickly determine if a specific location is listed on the various other components of the HSP. A variety of methods are used to implement improvements ranging from state or local forces performing the work to Federal-aid construction contracts.

A.3. Template #3: Internal Identification Processes

Overview

This approach to starting a LBR program leverages internal expertise within a transportation agency in identifying bottlenecks.

Major congestion issues are usually thoroughly identified by transportation agency personnel through the traditional regional and state long-range planning approach. However, there are perhaps hundreds of smaller scale targeted bottleneck problems which are overlooked simply because their scale does not compare to the larger congestion issues.

In this approach to setting up a LBR program, the districts and regions are called upon to identify these other smaller projects.

District and regional employees have the closest interaction with public; engage with them on a regular basis on a variety of traffic, congestion, or safety issues; and are in many times the best experts in identifying real bottlenecks that affect drivers every day.

Key Elements

- Low cost since agency staff are being polled and no public outreach is required; and
- Short timeline since many times, district and region engineers and planners already know where many local bottlenecks exist.

Timeframe	Cost	Institutional Hurdles	Analytical Complexity
Moderate/Long Term	Low	High	Moderate

Approach

Step TI. Establish the Types of Bottlenecks to Solicit

Agencies should first decide how broadly or narrowly they want to explore bottlenecks in their region. For example, a region could choose to only look for suggestions for pedestrian-related issues. Likewise, the focus could be on signalization issues and corridors or even centered on modal concerns.

Step 2. Determine Bottleneck Solicitation Mechanisms

Once the scope of the outreach is clarified, an approach to solicit the districts or regions should be developed. It is recommended that an internal "call for projects" be developed and disseminated to the districts and regions. This call should have explicit directions on the information required for each bottleneck to target. This information should include standard performance measures such as traffic and safety, but the submission criteria should be flexible enough to include nontraditional metrics. It is recommended that a relatively short timeframe be assigned for the projects to be submitted. A shorter timeframe will promote a "tiger team" approach and districts and regions will be more motivated to ensure projects in their region are included to compete with other statewide projects.

Automated survey mechanism (i.e., web-based) could be utilized as well but care should be taken to ensure they are robust enough to allow districts and regions flexibility in submitting a wide variety of projects.

Step 3. Develop Prioritization and Sorting Methodology

Once bottleneck locations are collected, a mechanism must be developed to rank and prioritize the responses. Transportation agencies routinely prioritize and rank projects. However, for this effort, it is not recommended that any traditional extensive modeling or benefit analyses be conducted. These traditional planning processes, although well established, are too cumbersome for this effort.

Instead, priorities should be established by agencies using more strategic and simplified ranking mechanisms. Examples include comparing the potential LBR projects with those in the TIP and LRTP or ranking projects higher if they have lesser environmental documentation.

The ranking could be conducted by engineers and planners from Central Office or Headquarters or an expert panel could be formed with representation from each district and region to determine the priorities of the bottlenecks.

Step 4. Develop Bottleneck Project Plans

Once projects are identified, project plans need to be developed and the projects programmed. It is recommended that a quick turn project plans be developed for these projects. Elements include geometric sketches, traffic and safety impacts, costs, and even rough benefit/cost analyses. The projects need to be developed so that they can either compete for set aside LBR dollars or compete with other projects in the traditional programming process.

Example

Mn/DOT was originally driven to explore low-cost congestion relief projects because of budgetary restrictions but quickly realized that these projects could be implemented very quickly (and thus were highly visible to the public). They also found that because of low costs, they could identify multiple locations all over the region – the projects could be "spread around."

The cornerstone of this process is the Congestion Management Planning Study (CMPS). This was developed as quick turnaround study so that projects could be recommended to the Legislature before it adjourned for the session. (The process was started in February and results achieved in May.) It was envisioned as a "tuning study" – how can the system be "tuned" in specific areas to get congestion relief rather than rebuilt. Although cast as a single study, it is hoped that it can be integrated as an ongoing process within the Department. The process works as follows:

- **Step 1: Project Identification.** Projects were identified from a number of different sources, including Mn/DOT project lists, Mn/DOT Area Managers, SRF Consulting Group, Inc., Mn/DOT Metro District's Safety Capacity, Mn/DOT's freeway congestion maps and the Governor's 2007 bonding list. By combining the information gathered from the sources, 184 projects were identified and included in this study as congestion management projects.
- **Step 2: Screening #1.** A series of binary tests (pass/fail) were applied to the projects, resulting in downsizing the list to 100 projects:
  – Project cost < $15M;
  – Project not in three-year TIP;
  – Could require project memorandum or lesser environmental documentation;
  – Annual hours of delay > 25,000 hours of congestion; – Freeway or Arterial > two hours of congestion; and

- – Arterial relieves parallel congested freeway or directly responsible for freeway congestion.
- **Step 3: Screening #2.** Qualitative criteria were applied, resulting in 60 projects still under consideration:
  - – Project implementation/design readiness; – Cost range;
  - – Congestion benefit (weighted delay);
  - – Traffic management for construction;
  - – Future demand changes; and
  - – Relieves congestion without adverse downstream affects.
- **Step 4: Expert Workshop.** Short-range congestion projects were prioritized by expert group during half-day workshop, resulting in 19 projects totaling $60.8M. The projects fell into three broad categories:
  - – Low-cost capacity improvements (e.g., auxiliary lanes);
  - – Restriping to change lane configuration (which the maintenance department could handle); and
  - – Traffic control device improvements (add ramp meters and "tune" signal timing).
- **Step 5: Project Planning.** For each of the 19 projects, the following project estimates were prepared:
  - – Geometric sketches;
  - – Type and scope of project;
  - – Congestion impacts;
  - – Safety impacts; and
  - – Estimated benefit-to-cost ratio.

Bottleneck Performance Measures

The State of Minnesota (not just the DOT) has been a leader in developing and using performance measures to assess how it is meeting customer expectations. In accordance with these principles, evaluation of completed projects is being done to track the effectiveness of current and future investments:

- Measurement of "before" and "after" project conditions to assess the project's effectiveness and build experience for the type of benefits those different projects can deliver;
- Annual system measures that can capture overall congestion trends for different systems over time (e.g., Texas Transportation Institute (TTI)

congestion index, percentage of Metro Urban Freeway System (MUFS) congested); and

- Measurement of strategies to shifts peak demands to off-peak periods (e.g., number of persons moved at the per lane capacity with speeds greater than 45 mph).

A preliminary list of performance measures being considered are as follows. Additional ones may be developed in the future and not all may apply to the bottleneck relief projects:

1) Percentage of miles congested (i.e., number of directional miles with speeds < 45 mph during peak periods);
2) Total daily delay (volume x time difference between actual and posted speed);
3) Reliability (TTI – Congestion Index/Travel Time Buffer Index);
4) Miles of FIRST/incident response coverage;
5) Number of hours where volume > capacity;
6) Frequency of signal retiming;
7) Percent of arterials with coordinated signals;
8) "Before/After" benefit/cost ratio of corridor improvements;
9) "Before/After" benefit/cost ratio of signal retiming;
10) Customer satisfaction survey of peak-hour travel (omnibus transportation survey);
11) Percent of MUFS instrumented;
12) Planned lane closures System;
13) Unplanned lane closures System;
14) Average clearance time for snow and ice removal;
15) Average clearance time for freeway incidents; and
16) Throughput (i.e., number of vehicles through a specific corridor or across a screenline over a specified time period).

Overall, the CMSP process follows a procedure to develop candidate projects. (Freeway Performance data are used in the screening but in a very high-level way). Mn/DOT's freeway data system provides lane by lane data for all controlled access road as well as all ramps – this allows them to easily created accurate and timely analysis on an as needed basis. This existing archive data system allowed Mn/DOT order to keep to the aggressive schedule. At the end of the process (Step 5) more quantification comes into play. This process worked extremely well in the eyes of Mn/DOT – they feel that a more labor intensive, and drawn-out procedure would have essentially yielded the same project list. The reason appears to be that current bottleneck problems are easily identified through visual inspection and via the freeway surveillance data, so a detailed analysis is not required.

Another aspect of the streamlining (in addition to the qualitative project identification and screening process) is that some larger environmental aspects could not be addressed (e.g., area drainage). In order to keep the costs low and the projects quickly implementable, these environmental concerns were thought to be within the scope of the longer-term "megaprojects" being planned for.

Bottleneck Funding

No separate funding program was established for the CMPS projects – they have to compete for funding with other projects. If projects are seeking Federal funding, then they have to go through the evaluation process established by the MPO, which has its own set of evaluation criteria.

Example Bottleneck Projects

Three low-cost bottleneck projects were highlighted by Mn/DOT. (These are being submitted as examples by the Division.) The positive impacts of these projects on congestion are dramatic. (See Table A.1 and A.2 for details.)

- **I-394 at Louisiana Street.** Mn/DOT added an auxiliary lane one mile long at a cost of \$2.6M. Previously, queues could back up for six miles on this section; after completion, queues were reduced to zero! (for recurring conditions).
- **I-94 in St. Paul.** A four-lane section of freeway connected to two six-lane sections (a lane-drop bottleneck). Queues were 2.0 miles in the eastbound direction. Mn/DOT increased the number of lanes to six throughout this extended segment at a cost of \$10.5M. Although the desired result was achieved, the existence of other major bottlenecks at the end of the segment (freeway-to-freeway interchanges) limited the effectiveness of this improvement.
- **TH-100 in St. Louis Park.** This is another example of a 6-4-6 lane drop with a highly restricted cloverleaf interchange in the four-lane section. The shoulder was used to provide an additional through lane; two close-spaced interchanges were connected with CD roads and the 1950s cloverleaf interchange was redesigned to a diamond interchange at a cost of \$7.5M. Northbound queues were reduced

from 5.25 miles to 0.25 miles and southbound queues were reduced from 6.0 miles to 0.25 miles.

Table A.1. Cost and Travel Time Benefit of Completed Congestion Management Projects[a]

	Project Cost (In Millions)	Reduction in Annual Hours of Delay	Estimated Annual Travel Time Benefit (In Millions)	Project Service Life (In Years)	Estimated Travel Time Benefit Over Project Service Life (In Millions)	Estimated Travel Time Benefit to Cost Ratio
I-394	$2.6	87,000	$1.1	20	$21.6	8
I-94	$10.5	139,000	$1.7	20	$34.6	3
TH 100	$7.1	1,063,000	$13.2	7	$92.3	13
Total	**$20.2**	**1,289,000**	**$16.0**	–	**$148.5**	–

[a] Congestion data for Tables A.1 and A.2 were provided by Mn/DOT Metro Traffic and based on freeway loop detector data from 2004 to 2007.

Table A.2. Other Benefits of Completed Congestion Management Projects

	Peak Period Vehicle Flow Increase	Decrease in Miles of Congestion	Increase in Peak Period Speeds	Preliminary Safety Impacts
I-394	4,650	6.0	30 mph in p.m.	Sixty percent reduction of property damage crashes, no change in injury crash rate.
I-94	3,200	2.5	40 mph in a.m. 25 mph in p.m.	Modest reduction in number of mainlinecrashes.
TH 100	14,450	10.75	45 mph in a.m. 30 mph in p.m.	Thirty percent reduction of property damage crashes, 70 percent reduction of injury crashes.

End Notes

1 Weiner, Edward, Urban Transportation Planning in the United States: An Historical Overview. Revised and Expanded Edition. Praeger Publishers, Westport, Connecticut, 1999.

2 Washington State Department of Transportation, The 2010 Congestion Report, November 2010, http://www.wsdot.wa.gov/Accountability/Congestion/2010.htm.

3 http://www.wsdot.wa.gov/planning/HSP.

4 http://www.wsdot.wa.gov/Accountability/PerformanceReporting/GMAP.htm.

INDEX

N

O

P

Q